New Knowledge Creation Through ICT Dynamic Capability

Creating Knowledge Communities Using Broadband

New Knowledge Creation Through ICT Dynamic Capability

Creating Knowledge Communities Using Broadband

By

Mitsuru Kodama

INFORMATION AGE PUBLISHING, INC.
Charlotte, NC • www.infoagepub.com

Library of Congress Cataloging-in-Publication Data

Kodama, Mitsuru, 1957-
  New knowledge creation through ICT dynamic capability creating knowledge
communities using broadband / by Mitsuru Kodama.
      p. cm.
  Includes bibliographical references.
  ISBN-13: 978-1-59311-874-7 (pbk.)
  ISBN-13: 978-1-59311-875-4 (hardcover) 1. Knowledge management. 2. Broadband
communication systems. I. Title.
  HD30.2.K6397 2007
  658.4'038–dc22

                                                                    2007047724

CONTENTS

List of Figures ix

Preface and Acknowledgments xiii

1 New Knowledge Creation Using Broadband Networks 1
Business Innovation Using Broadband 1
ICT Technology and Market Innovation 4
   The Hop Process (From the 1950s to the Mid-1980s) 5
   The Step Process (Mid-1980s to Mid-1990s) 6
   The Jump Process (From the Mid-1990s to the Present) 8
The VIN Concept 9
   Structure of Use for Interactive Real-Time Models 11
   Structure of Use for Non-Real-Time Models (Storage System) 12
VIN Tools as Network Strategy Support Tools 13
VIN Tools as a Design for Ba and Communities of Practice 18
Acquiring ICT Application Capabilities 20
The Enactment of New Business Structures Through VIN Tools . . 22
New Knowledge Creation Through ICT Dynamic Capability 26
Fusion with Wireless Broadband 27
The Structure of This Book 30
Notes 32

2 Knowledge Creation Process Through ICT Dynamic Capability:
Theoretical Framework 33
Competitive Advantage Through ICT: What Is ICT Capability? . . . 34
New Knowledge Creation Through ICT: Theoretical Framework . 37
   The Significance and Essence of Knowledge Management
      for the ICT Era 37
   Knowledge Creation Process Through the SECI Model 38
   Empirical Background of the SECI Model 41

Knowledge Creation Model Through Knowledge-Handling Process . . . 42
Community Knowledge Creating Cycle in Real and Virtual Space:
　　Theoretical Framework 43
Three Layers in Knowledge Boundaries 45
The Community Knowledge Creating Cycle Resulting
　　from Communities of Practice 47
The Community Knowledge Creating Cycle Resulting
　　from Strategic Communities 48
ICT Dynamic Capability: The Theoretical Framework 51
　ICT Application Capability 52
　Architect Capability of Context 54
　Boundary Consolidation Capability 56
Notes 58

3 Network Strategy As Practice 59
　VIN Tools as Network Strategies 59
　Boundary Communications 62
　Boundary Communications and Strategy-Making Processes 63
　　The Top-Down Approach (Pattern 1) 64
　　The Bottom-Up Approach (Pattern 2) 66
　　Shifting from a Bottom-Up to a Top-Down Approach (Pattern 3) 67
　　Shifting from a Top-Down to a Bottom-Up Approach (Pattern 4) 68
　　Combining a Top-Down and Bottom-Up Approach (Pattern 5) 69
　The ICT-Driven Community-Based Firm 71

4 Promoting Community Management through VIN 75
　The Framework of Community Management 76
　　Community Knowledge and Innovative Leadership
　　　by Community Leaders 77
　　Promoting Community Management 78
　VIN in the World of Finance 80
　　VIN within and between Companies: Application
　　　of Type 1 and Type 2 Communities 81
　　Dealing and Trading: Type 3 Community Application—
　　　Specific Customers 83
　　Multimedia Banking Kiosks: Mass Users—A Type 3
　　　Community Application 83
　　The World's First Mobile Videophone "Visual Call Center" 84
　　A New, Integrated VIN for Finance Businesses 86
　VIN in the World of Automobiles: Managing with Speed
　　and Excellence—the Case of Peregrine, Inc. 89
　　Reforming to Become a World-Class Company
　　　in Ten Years via VIN 91
　　Top-Down Investment in ICT 93
　Building ICT Dynamic Capability Through a Top-Down
　　Approach 93
　Notes 95

5 The Promotion of Strategic Community Management
Utilizing Video-Based Information Networks 97
Introduction. 98
Strategic Community Creation and VIN 99
    The Organizational Learning Support of Strategic
        Communities Using VIN. 100
Case Study: Innovation in the Field of Veterinary Medicine
        Using VIN 102
    Creation of Innovation by Merging Different
        Business Areas 104
    Bottom-Up Strategy for ICT Solution Proposals
        to the Customer: Implementing Emergent Strategies 105
    Developing and Constructing the New VIN 107
    Innovation in Community Knowledge through VIN 109
    The Strategic Community Startup Phase (see Figure 5.5) 110
    The Strategic Community Growth Phase (see Figure 5.5) 112
    The Strategic Community Development Phase
        (see Figure 5.5) 113
    The Community Competence Sophistication Process 114
    Interactive Linkage between Community Knowledge
        and Community Competence. 116
    ICT Dynamic Capability by Strategic Community. 118
    The Superiority of VIN 121
    Conclusion. 122
Notes. 122

6 Enabling Emergent and Deliberate ICT Strategies. 125
Customer Value Creation Management (IBIZA, Inc.) 126
    Past Issues. 127
Productive Interpretation and Creative Realization 127
ICT Investment for the Customer 129
VIN Supporting Customer Value Creation Model Businesses
        (Type 2 and Type 3 Communities Application
        in Figure 4.1) 129
Continual Sharing, Inspiring, and Creating Community
        Knowledge (Type 1 Community Application
        in Figure 4.1) 130
    Communication and Collaboration Using High-Quality Video
        Information 131
    On-Demand Searches of Shop and New Product Information 132
Customer Value Creation with the Customer through Mobile
        Phone Networks 133
Creating "ICT Dynamic Capability" through Different
        Strategy-Making Processes 134
Notes. 138

7   Managing Strategic Management Cycles: A Case Study of Sony. . .141
Sony Marketing's ICT Strategy: A Case Study 141
ICT Dynamic Capability and New Knowledge Creation 146
The ICT Dynamic Capability Loop and the Community
    Knowledge Creating Cycle 147

8   Managing Paradox Through Dialectical Management 151
Broadband IP Videoconferencing Initiatives: A Case Study
    from Otsuka Corporation. 152
    Step 1 Initiatives. 152
    Step 2 Initiatives. 153
    Step 3 Initiatives. 154
Initiatives to Change Corporate Culture: A Case Study
    from NTT. 157
    Construction of New Organization by Top Management Members . . .157
    Development of Yarima SHOW Multistrategy. 159
    Three Restrictions Lifted by ICT Tools 160
    Meeting of 100,000 People 162
Integrating Different Strategy Perspectives 163
Building a Platform of Resonating Values via Dialectical
    Management 165
Enhancing ICT Dynamic Capability through Spiral Resonance
    of Value 170
Notes. 171

9   New Knowledge Creation Through ICT Dynamic Capability. 173
Deep Collaboration, Dialectical Dialogue, Resonance of Value,
    and Trust Building within Knowledge Communities 174
    Innovative Leadership by Management Leaders. 177
    Value-Based Leadership 178
    Dialectical Leadership. 180
Network Collaboration-Based Organizations. 186
    Implementing Strategy in Time and Space through ICT. 187
    Network Collaboration-Based Organizations and Networked
        Knowledge Communities 190
    Building Network Collaboration-Based Organizations. 193

References. 197

LIST OF FIGURES

Figure 1.1	Global Trends in Broadband User Numbers	2
Figure 1.2	Global Cell Phone Distribution	3
Figure 1.3	Social Impact of Broadband Technology	3
Figure 1.4	A Grand Design for Organizational Management and the Spread of ICT	5
Figure 1.5	Example of Knowledge Management Using ICT	7
Figure 1.6	Expanding the Visual Communications Market	10
Figure 1.7	Technological Innovation in VIN from Advances in IP and Semiconductors	10
Figure 1.8	Video-Based Information Network (VIN) Concept	11
Figure 1.9	Example of PC-Type Videoconferencing System	12
Figure 1.10	Positioning of VIN	13
Figure 1.11	VIN Tools as Management Support Tools	14
Figure 1.12	Practice of Knowledge Management with VIN Tools	15
Figure 1.13	VIN as a Third Party–Oriented Communication Tool	16
Figure 1.14	Setting and Design of All Types of Meetings as Ba and Community of Practice (Example of C Company)	19
Figure 1.15	New Media Literacy for VIN Tools	21
Figure 1.16	New Application Capabilities for VIN Tools	22
Figure 1.17	Usage Structure of VIN Tools	23
Figure 1.18	Examples of VIN Tools in Business	24
Figure 1.19	Examples of VIN Tools in Medicine and Welfare	24
Figure 1.20	Applicable Domains for VIN Tools	25
Figure 1.21	Creating Communities with VIN Tools	25
Figure 1.22	Dynamic Loop for ICT Activity	28

Figure 1.23   FMC Through VIN Tools. 29

Figure 1.24   Broadband Communications between Mobile
and Fixed Videophones. 29

Figure 1.25   VIN Tool Application Methods Among Broadband
Mobile Devices 30

Figure 2.1   Process of SECI Model. 39

Figure 2.2   New Knowledge Creation through Community
Knowledge Creating Cycle. 44

Figure 2.3   Community of Practice and Strategic Community (SC) . . . 45

Figure 2.4   Interaction Between Real Space and Virtual Space:
Case of New Product Development 55

Figure 2.5   Conceptual Framework of ICT Dynamic Capability 57

Figure 3.1   Superiority of VIN Tools Over Groupware and Email. 60

Figure 3.2   ICT Dynamic Capability Through Network Strategy. 62

Figure 3.3   ICT-Driven Community-Based Firm 72

Figure 4.1   Community Types. 76

Figure 4.2   Spiral Business Innovation through Community
Management. 79

Figure 4.3   Creating Customer Value Through Visual Call Centers . . . 85

Figure 4.4   Image Guidance Function of Visual Call Center 86

Figure 4.5   New Integrated VIN for the Finance Industry 88

Figure 4.6   The VIN at Peregrine, Inc. 92

Figure 5.1   Strategic Community Management 101

Figure 5.2   Innovation of Spiraling Community Knowledge-
Creating Cycle via Video-Nets 102

Figure 5.3   Telemedicine and Distance Learning Systems
That Were Developed and Their Manner of Use. 108

Figure 5.4   Process of Innovating Community Knowledge 109

Figure 5.5   Community Knowledge-Creating Cycle 110

Figure 5.6   Nationwide Expansion of Communities. 111

Figure 5.7   Advanced Process of Community Competence 115

Figure 5.8   Linkage between Community Knowledge
and Community Competence. 117

Figure 5.9   ICT Dynamic Capability and New Knowledge Creation . . 119

Figure 5.10   Dynamic Loop of ICT Activity. 120

Figure 5.11   Superiority of VIN Over Groupware and Email. 121

Figure 6.1   IBIZA's Vertically Integrated Management 126

Figure 6.2   IBIZA's VIN Network Organization 131

Figure 6.3   Community Knowledge Creation by VIN at IBIZA, Inc. . . . 132

Figure 6.4   Forging Closer Bonds with the Customer by Connecting
via Mobile Phones 133

Figure 6.5   ICT Dynamic Capability and New Knowledge Creation . . . 135

Figure 6.6   IBIZA's ICT Dynamic Capability. 138

Figure 7.1 Rise in VIN Operating Ratio Through Emergent
Strategy. 144

Figure 7.2 ICT Dynamic Capability and New Knowledge Creation . . 146

Figure 7.3 ICT Dynamic Capability Loop 148

Figure 7.4 The Relationship Between the SICA Model and the Three
Elements of ICT Dynamic Capability 149

Figure 7.5 Relationship Between ICT Dynamic Capability Loop
and Community Knowledge-Creating Cycle 150

Figure 8.1 Osaka Corporation's VIN (Step 3) 156

Figure 8.2 ICT Tools for Knowledge Management through
Yarima SHOW Multi Strategy 160

Figure 8.3 Leaders' Communities and Strategy Structures
for Each Management Layer: Case Study from Otsuka
Corporation 164

Figure 8.4 Leaders' Communities and Strategy Structures
for each Management Layer: Case Study from NTT. 164

Figure 8.5 Resonance Process of Value in the Community
through Dialectical Management—Unification
and Conflict of Opposites 166

Figure 8.6 Structure of Platform for Resonating Values
through Dialectical Management and ICT Application
Capability 171

Figure 9.1 Deep Collaboration, Dialectical Dialogue, Resonance
of Value, and Building Trust 177

Figure 9.2 Innovative Leadership by Management Leaders 178

Figure 9.3 Dialectical Management Realizing "Space Strategy". 185

Figure 9.4 Features of Network Collaboration-Based
Organizations. 186

Figure 9.5 Network Collaboration Management on a Space Axis . . . 188

Figure 9.6 Network Collaboration Management on a Time Axis. . . . 189

Figure 9.7 ICT Strategy on a Time Axis 190

Figure 9.8 Networked Knowledge Communities. 191

Figure 9.9 Network Collaboration-Based Organizations 192

PREFACE AND ACKNOWLEDGMENTS

Nowadays the world is endlessly creating new knowledge from connections among people, organizations, and companies in the virtual space of the broadband network. The music distribution business (centered on the iPod), the Google and Yahoo! search businesses, the various broadband services, and the dramatic global spread of mobile phones, bring great changes not just to our lifestyles, but also to company business styles. ICT has been developing rapidly for the last decade as a core infrastructure for diverse broadband network services. Fixed phone services as voice transmissions, which have come to be dominated by fixed telecommunications networks for around 100 years, are promoting the next-generation network worldwide as IP-based multimedia network technology with completely integrated data, voice, and video. Japan is currently the world's most advanced country for developing and installing fiber optics communications, and is taking a global lead in this technology field. Moreover, Japan is developing and commercializing its "i-mode" mobile Internet service (mobile phone technology) as a global pioneer, and is taking this mobile phone service technology from the third to the fourth generation.

One topic that should be noted concerning this rapidly developing ICT is the technological innovation that features broadband with "video communications" at its core. I have come to pursue many ICT developments in the past, as well as the e-business service development applying these commercializations, installations, and technologies. Then in this book, especially among these areas, I would like to analyze and consider all new knowledge creation, strategy, capability, organization, leadership, and process points with regard to video-based technological innovation and the corporate change and service innovations that apply this technology. Regarding the book's contents, I have spent the past 17 years analyzing and

New Knowledge Creation Through ICT Dynamic Capability, pages xiii–xiv
Copyright © 2008 by Information Age Publishing

considering accumulated research from the standpoint of information and management studies with regard to "video communications technology and application." This book's greatest aim is to bridge ICT-activated management theory and actual business practice. Put another way, my greatest research objective is to build a dynamic practical management theory that supports the daily or hourly changing business environment.

The need to dynamically grasp the phenomena of business activities led me to follow individual business cases in detail and gather chronological data for content emphasizing field studies. As a result, my research methodology has derived a core theoretical framework from this gathered data, based on grounded theory and focused on ethnography, participant observation, and in-depth interviews at the workplace.

The completion of this book is the fruit of interaction and collaboration with numerous practitioners. At this juncture, I would like to express my gratitude to several of them, especially for the many suggestions received from Mitsuo Kanebako, managing director of NTT BizLink Inc. I would also like to express my thanks to my wife and children for their energy and encouragement every day during the creation of this book. Finally, I would like to express my deep appreciation to George Johnson, president of Information Age Publishing, for giving me the precious chance to publish this book.

Finally, I have used fragments of material I have previously published. These are refereed papers cited below. I am deeply appreciative to the following publishers for their permission to reuse this material. Permission has been received from Elsevier Science to use the following papers:

Kodama, M. (1999). Customer value creation through community-based information networks. *International Journal of Information Management, 19*(6).

Permission has been received from Emerald to use the following papers:

Kodama, M. (1999). Community management support through community-based information networks. *Information Management and Computer Security, 7*(3).

Kodama, M. (1999). Strategic business applications and new virtual knowledge-based businesses through community-based information networks. *Information Management and Computer Security, 7*(4).

Kodama, M. (2002). The promotion of strategic community management utilizing video-based information networks. *Business Process Management Journal, 8*(5).

CHAPTER 1

NEW KNOWLEDGE CREATION USING BROADBAND NETWORKS

BUSINESS INNOVATION USING BROADBAND

In recent years, new businesses have been created as communications functions have grown more advanced through broadband networks, exemplified by ICT (information and communication technology). Broadband networks are not just an advance in information processing centered on computers using conventional IT. They are technological infrastructures that enable interactive networking of individuals all over the world, and the transmission, sharing, and creation of diverse data, information, and knowledge held by individuals.

The number of broadband users worldwide broke through the 100 million mark in 2003. Broadband is expanding rapidly in Asia and Europe, fast on the heels of North America, and rapid growth is anticipated for the future. Household broadband is around 20% in Europe (see Figure 1.1). Broadband distribution is highest is South Korea, followed by Japan, where the takeup is accelerating. Japan, China, and South Korea together have a total of 45 million subscribers, and the distribution is expanding. Among these broadband types, FTTH (fiber to the home) is gaining attention as the ultimate broadband technology. FTTH reached 10 million subscribers in Japan as of the end of March 2007, putting Japan first in terms of subscriber numbers to the new technology. Meanwhile, Europe and U.S. broadband technologies mainly focus on DSL or CATV.

New Knowledge Creation Through ICT Dynamic Capability, pages 1–32
Copyright © 2008 by Information Age Publishing

1

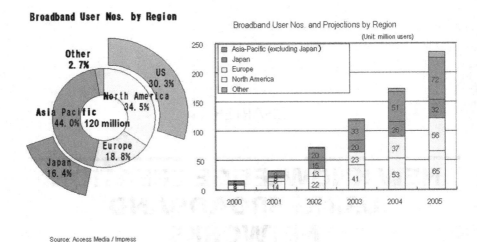

Figure 1.1. Global Trends in Broadband User Numbers.

As for mobile phones, distribution rates have already topped 50% in the major user countries (see Figure 1.2), and there is further room for growth in countries such as China, India, and Brazil, where the distribution rate is still low. Japan and South Korea have the highest mobile Internet support rates for technologies, led by i-mode and others, and new consumer services and businesses using mobile phones are growing. The 3G technologies that are the gateway to broadband, especially for mobile phones, are spreading, especially in Japan but also in other Asian countries and Europe.

Progress of this kind of broadband technology is having a big impact on individual lifestyles and corporate activities (see Figure 1.3). Rich broadband services have the potential to deliver rich lifestyles by offering individual comfort, convenience, and security as well as a varied lifestyle environment. For corporate activities, too, broadband use goes beyond improving management efficiency to contributing to enhancing customer services and developing new markets. The shape of corporate organizations and behavior is changing along with recent changes in the business environment and development of broadband networks. It will become increasingly important for future business strategies to go beyond resources limited by business units within conventional corporate organizations to take positive initiatives with personnel and knowledge outside the company as well as with the dynamic use of ICT, through such means as external strategic alliances, virtual corporations, mergers and acquisitions, and outsourcing (Kodama, 1999a, 1999b).

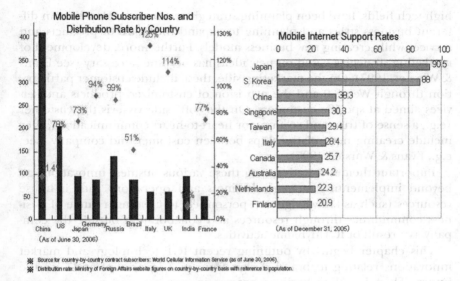

Figure 1.2.   Global Cell Phone Distribution

Figure 1.3.   Social Impact of Broadband Technology

For several years, various cutting-edge companies with core competences including software, semiconductors, and networking technologies forming the basis of broadband networks, as well as venture companies in high-tech fields, have been planning strategic business development in different business styles and planning to expand the share of products and services while creating new business models. Furthermore, development of marketing strategies based on new ideas has become necessary (see Evans & Wurster, 1997). On the marketing side, these include customer participation through Web 2.0 and the provision of customized products and services aimed at specific customers. On the"soft" side vis-à-vis the customer (e.g., a sense of trust and security for heart-to-heart communication), they include creating new relationships between customer and company (see, e.g., Evans & Wurster, 1997).

Important themes for advancing these various business innovations go beyond implementing strategic business and operations with in-house resources (such as knowledge and personnel) to creating a range of business communities through resources and collaboration outside the company as a result of ICT dynamic activities.

This chapter begins by outlining recent ICT technology and market innovations relating to brand design from past to present and into the future. Then I outline business process innovations and new e-business that activates the concepts and technology of video-based information networks (VIN) based on the leading broadband technologies of multimedia communication networks. Next, I touch on the possibilities of maximizing VIN to support speed and excellence in corporate management while simultaneously generating e-business that creates customer value. Dynamic activities with VIN as a future network strategy support tool will enable business innovation through strengthening corporate competitiveness and enhancing customer services.

ICT TECHNOLOGY AND MARKET INNOVATION

Amid a frenetic business environment, dynamic ICT applications are essential to steer corporate management in a more strategic direction and to raise business efficiency and productivity while enhancing management levels and organizational activity (see, e.g., Ross, Beath, & Goodfue, 1996; Venkatraman, 1991, 1994, 1997). Recent e-business through broadband and the greater efficiency of business processes and operations both within and among businesses are leading examples of this activity. With regard to these corporate organizations, I would now like to outline the grand design of organizational management with past, present, and future perspectives, and the ICT that has a great influence on this management (see Figure 1.4).

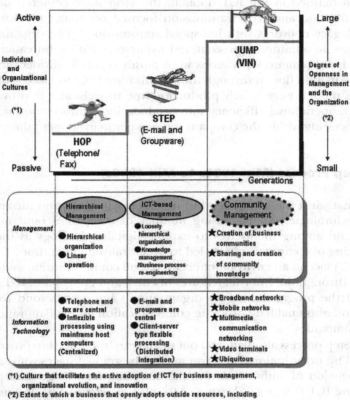

(*1) Culture that facilitates the active adoption of ICT for business management,
organizational evolution, and innovation
(*2) Extent to which a business that openly adopts outside resources, including
customers, is developed

Source: Kodama (1999b), modified

Figure 1.4. A Grand Design for Organizational Management and the Spread of ICT

The Hop Process (From the 1950s to the Mid-1980s)

In an era when international digital networks, such as today's broadband, were still undeveloped, most corporate organizations relied on the telephone and fax to carry out daily operations within and outside the company, whether with its own divisions or with clients and customers. All kinds of business processing took place centered on intensive, fixed processing actions typified by calculations from large, general-purpose host computers. This process is defined as the Hop stage of the time axis (see, e.g., Nolan, 1979).

When defined in the ICT domain, this Hop stage centered on voice communication and data transmission focused on analog networks and digital legacy networks with low-speed transmissions. The organizational shape was an administration-centered pyramid based on bureaucratically layered management. Operations were mostly vertically divided line-type, and information flowed through the organization from top to bottom. The business models were mostly production-type models, and improving the accuracy, operational efficiency, and productivity of information transmission processing within the company was a major management theme.

The Step Process (Mid-1980s to Mid-1990s)

Later, against the background of developing Internet and intranet network technologies, an increasing number of companies built networks within and among themselves to exploit digital technology as the huge downsizing of computer systems led to decentralized integration. This process took place in an environment centered on communication and collaboration through the Internet features of email and groupware. ICT activity enabled the progression of an organization's business beyond the Hop stage, and also enabled corporate communication and collaboration with other companies.

The Step process also flattened out the hierarchy of the Hop process, and speeded up organizational decision making by gradual stages while advancing delegation of authority. It was the era when knowledge management exploiting ICT (see, e.g., Skyrrme, 2001, also see Figure 1.5), business process reengineering (Hammer & Champy, 1993), and business process management through ERP (enterprise resource planning) began to be applied within the company. Thus it was a time when the importance of ICT-based management slowly permeated at the corporate level. It was also a time when the concepts of network organization as a corporate concept (Nohria & Ghoshal, 1997), the virtual corporation (Davidow & Malone, 1992), and the virtual team (Lipnack & Stamps, 1997) began to attract academic interest.

Meanwhile, it became clear that best practice regarding knowledge management is not necessarily applicable to all organizations and companies (see, e.g., Davenport & Prusak, 2000). The reason for this is that the matching of ICT with organizational landscape and context is key to the success of introducing ICT. While ICT was an enabling factor for changes in organizational structure, it also became clear that the kind of changes to bring about in the organization was determined by organizational purpose and context (see, e.g., Malone & Crowston, 1994). Accordingly, a large number of businesses recognized the importance of simultaneously implementing changes in personnel, organization, and corporate culture for

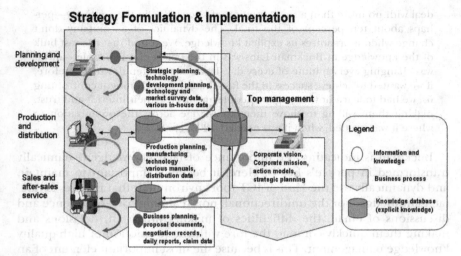

Figure 1.5.   Example of Knowledge Management Using ICT

business process innovations through ICT (Davenport, 1993). While numerous organizational themes remain, however, this kind of decentralized ICT environment has come to be recognized by a large number of companies as infrastructure that clearly promotes information transmission and sharing for actors within and outside the company, and enables expectations for individual and organizational activity.

Today, almost all companies promote communication and information sharing among organizations by incorporating email and all kinds of groupware. As a result, a lot of companies are also aiming to further enhance individually held information and knowledge as a goal of corporate innovation. Companies incorporating these new tools that are already undergoing systemic change have come to face the fact, however, that email groupware alone cannot sufficiently strengthen communication or speed up decision making inside and outside the organization.

Actors exploit the ICT tools of email and groupware, and promote the sharing and transmission of explicit knowledge specified as data (Nonaka & Takeuchi, 1995). Boosting the efficiency of corporate business processes through sharing of explicit knowledge is certainly possible. Steven Buckman, CEO of US company Buckman Laboratories, who is known for promoting knowledge management with ICT, commented as follows on knowledge essential for an organization:

> This does help, at least initially, by causing knowledge that has been written down to be organized and made available to the organization. Useful as that step is, however, we've found that it is not sufficient to achieve success. It can

deal with no more than a small fraction of knowledge in the company—perhaps about ten percent of the total. The dynamics of a company don't change when it organizes its explicit knowledge. We found that the vast bulk of the knowledge in Buckman Labs was in the heads of our people—and it was changing every minute of every day. It was not written down. Therefore, if we wanted to achieve success in the fast-changing environment confronting us, we had to learn how to engage people and arouse their interest and trust, making them willing to move their knowledge across the organization to where it was needed, when it was needed. (2003, p. 20)

Buckman is indicating the importance of tacit knowledge dynamically transformed in people's heads. Here, it becomes important to maintain and dynamically use the kind of ICT tool environment that inspires actors' tacit knowledge. For the unidirectional, non-real-time correspondence and discussions of email, the difficulties of making important decisions and making them quickly come to the fore with the promotion of high-quality knowledge management. This is because the most important element of an organization's communication is the empathy and solidarity between individuals arising from face-to-face communication. In this area, a sense of obstruction arises where email and groupware cannot communicate individual thoughts and feelings sufficiently. The ultimate communication leading to management innovations comes from the Jump process.

The Jump Process (From the Mid-1990s to the Present)

As mentioned above, the pattern that can be expected to grow increasingly as the future shape of business is business development fusing various business communities within and outside the corporate organization that openly incorporates internal and external resources and knowledge, including the customer (see, e.g., Chesbrough, 2003; Kodama, 2007a). This process creates a flatter, more flexible organizational structure and more open corporate management than does the Step process.

With the Jump process, moreover, community management (Kodama, 1999a) dynamically driving business strategy both within and among business communities (including customers) through the leadership of top and middle management becomes important. Network strategy tools to support the leaders that promote community management also become important.

Specifically, these are VIN tools that support business and the sharing and creation of knowledge and competence for management leaders. They are based on multimedia communications that prioritize the high-quality communication and collaboration involved in management leaders' decision making and interactivity among individuals. The essential points of this Jump step are already spreading around the globe. This is due to the positive incorporation of VIN tools based on dynamically innovating individual cultures and entire organizational landscapes as well as on multimedia communication networking. It is becoming important to drive business dynamically by opening up management and organizations through diverse business communities operating within and outside the company.

THE VIN CONCEPT

The market for the multimedia tools of videoconferencing systems, videophones, web meetings, and mobile videophones (these systems are referred to as "video terminals" in this text) is growing worldwide against a backdrop of broadband development. Video terminals gradually entered the boardrooms on the coattails of the dramatic spread of ISDN (Integrated Service Digital Network), centered on Japan and Germany, from 1995 onward. The distribution of these terminals has accelerated over the past several years with the move to broadband and IP (Internet Protocol) taking place alongside falling prices. Key features are the introduction of IP videoconferencing for new users and the videoconference user transferring to IP through ISDN based on conventional switching circuits. In the future there is a high possibility that the further spread of broadband and IP will further reduce product and communication costs, and that this will encourage user expansion and lead to still further cost reductions, thus growing the market and precipitating the positive feedback of network externalitics (Shapiro & Varian, 1998). (See Figure 1.6.)

Up until now, development vendors' technology themes for ISDN and dedicated circuit videoconferencing systems have focused on high-quality film and voice compression based on switching circuit technology. In this area, reliability and stability were always required from system development, and it became necessary to pursue sustaining technology (Christensen, 1997) capable of efficiently compressing and transmitting high-quality multimedia data over narrow bandwidths. The high cost of the systems, together with communication costs and volume restraints, posed great hurdles for users wanting to introduce the system.

Thanks to the development of the "disruptive technology" (Christensen, 1997) of IP and mobile videophones, videoconferencing systems appeared that were still faster, cheaper (fixed price systems for always-on connections), and better quality than the high-functioning, high-priced systems that had gone before (see Figure 1.7). Moreover, the hardware-based product groups of the ISDN era diversified as developments in software-focused products paralleled those in IP and in semiconductor speed and functionality.

The development of IP for video terminals enabled systems to offer much more than the conventional full-scale videoconference based in the

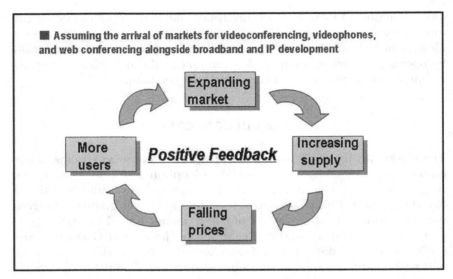

Figure 1.6.   Expanding the Visual Communications Market

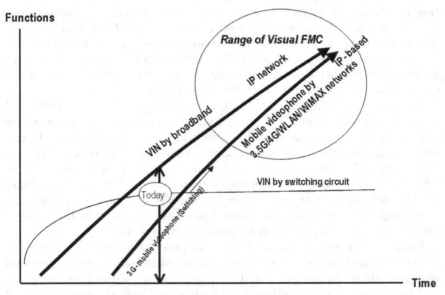

Figure 1.7.   Technological Innovation in VIN from Advances in IP and Semiconductors

meeting room. The range of systems diversified in response to users aims to include PC videoconferencing systems for groupware, collaboration, and chat as well as integrated videophones. Multipoint connection units

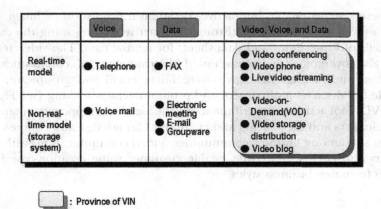

	Voice	Data	Video, Voice, and Data
Real-time model	● Telephone	● FAX	● Video conferencing ● Video phone ● Live video streaming
Non-real-time model (storage system)	● Voice mail	● Electronic meeting ● E-mail ● Groupware	● Video-on-Demand(VOD) ● Video storage distribution ● Video blog

▢ : Province of VIN

Source: Kodama (1999b), modified

Figure 1.8. Video-Based Information Network (VIN) Concept

based on low-priced software promoted independent use by companies, while companies that prioritized the reliability and quality of IP networks made positive use of multipoint connections from outsourcing services provided by ASPs (Application Service Providers), leading the ASP market to grow worldwide.

The VIN tools that exploit video terminals in this way are networks that speed up individual communication and collaboration by various text-based means, led by email and groupware, and enable interactive communication and collaboration integrating the three multimedia elements of image, voice, and text. These VIN tools are already realizing broadband technology development and technology innovations in a wide range of network devices. The usage models are interactive real time and non-real time as a storage system (see Figure 1.8).

Structure of Use for Interactive Real-Time Models

As shown in Figure 1.6, the first structure of use is an interactive real-time system is to exchange information at the same time between different places. This leading structure encompasses personal desktop videoconference systems (DTC), dedicated room system videoconferencing models, business-oriented web conferencing systems, simple videophone software for use on PCs and simple videophones targeting the general user (examples such as the free Skype service are widespread among videophones), and video terminals, especially mobile videophones on 3G handsets. Video streaming of live images distributed to multiple PCs and mobile phones can also be considered as a real-time model of a VIN tool in the wide sense.

These systems enable interactive, real-time transmission of image and voice as well as collaboration through interactive sharing using the essential tools of PowerPoint and data sheets for textual data. The video terminals also overcome the often noted disadvantages of communicating opinions and feelings through non-real-time email and groupware, and enable full discussion of content and prompt decision making (see Figure 1.9). VIN tool activities contribute to decision-making support for management leaders and the sharing and creation of knowledge and competence within and among business communities. Video communications with customers and clients, moreover, enable customer value creation-type businesses from new business styles.

Structure of Use for Non-Real-Time Models (Storage System)

The second structure of use is a cumulative system that searches for and extracts knowledge and competence on-demand and nonsimultaneously. Video-on-demand (VOD) is a leading model for this system. VOD enables access to and search of system image (including voice and text information) databases using PC, mobile phone, and videoconferencing systems as well as videophones and other video terminals.

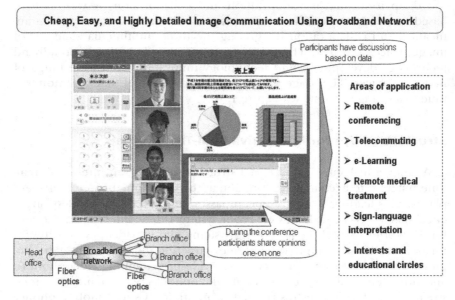

Figure 1.9. Example of PC-Type Videoconferencing System

VOD is a system that stores, in image form, knowledge and competence that is difficult to communicate by text, and enables sharing within and among the business community. It is already being realized as a major achievement of e-business aimed especially in research and development, design and production, and personnel education fields such as staff training as well as at the consumer. Video storage and delivery systems aimed at the general consumer through YouTube and video blogs also belong to the category of non-real-time VIN tools in the wider sense of the term.

VIN TOOLS AS NETWORK STRATEGY SUPPORT TOOLS

Looking back on ICT development, the large-scale host computer as an unchanging data system was positioned as a business management system based on numerical information processing within routine operations work. Meanwhile, email and groupware promoted information sharing among actors pooling data as explicit knowledge inside and outside the company. I see VIN as strategic network support tools with a stronger leaning toward management innovation and customer value creation models than email and groupware (see Figure 1.10).

Nissan Motor Company CEO Toshiyuki Shiga (2006) describes the effects of VIN tools as follows:

> The videoconferencing system is highly useful from a time-saving perspective. Nissan bases are spread out all over the world, and we have frequent discussions with major shareholder Renault of France, so we are often active in

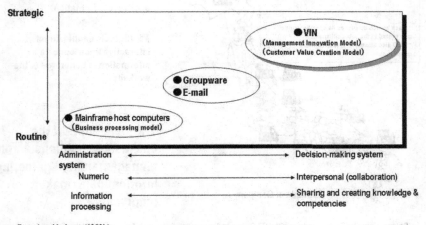

Source: Based on Kodama (1999b).

Figure 1.10. Positioning of VIN

videoconferencing on our PC screens. In my case, I use it once or twice a day for an hour or so each time. It is still important to walk around the worksite, but in some situations its use can boost productivity a great deal. Image movement used to be clumsy, and it was difficult to read a partner's expression. The current system has advanced, however, and you can speak with the sense that they are in the same room as you. You can communicate not just words but also body language through gestures, enabling rich communication. There is no sense of disharmony with videoconferencing, even during delicate negotiations.

The essence of VIN tool activities falls into three categories. The first is the achievement of speed management. VIN tools support information gathering and faster decision making among management leaders (see Figure 1.11). Top managers can communicate their intentions interactively to all employees while providing all data directly. Top management can, moreover, be acquainted with live data directly and interactively from the place where it is generated. VIN tools are also strategic tools for the promotion of knowledge management and the delivery and sharing among all employees of information and knowledge held not just by management leaders but by large numbers of staff (Figure 1.12).

A businessperson, for example, communicates management results or a progress report together with its spoken nuances, and businesspeople in various regions are able to instantly share that deep information or knowledge.

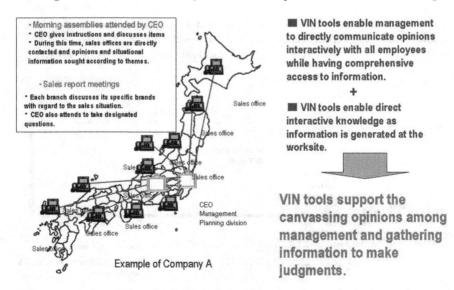

- Morning assemblies attended by CEO
* CEO gives instructions and discusses items
* During this time, sales offices are directly contacted and opinions and situational information sought according to themes.

- Sales report meetings
* Each branch discusses its specific brands with regard to the sales situation.
* CEO also attends to take designated questions.

Sales office

CEO
Management
Planning division

Example of Company A

■ VIN tools enable management to directly communicate opinions interactively with all employees while having comprehensive access to information.
+
■ VIN tools enable direct interactive knowledge as information is generated at the worksite.

VIN tools support the canvassing opinions among management and gathering information to make judgments.

Source: Compiled from interviews with NTT Bizlink and presentation materials for use outside the department.

Figure 1.11.   VIN Tools as Management Support Tools

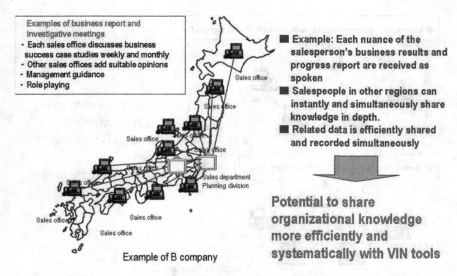

Figure 1.12.   Practice of Knowledge Management with VIN Tools

VIN tools also possess the functions of encouraging management and leaders to share information and context with third parties. Put another way, it is a third-party-oriented communication tool (see Figure 1.13).

The second category is raising management efficiency, an area where innovation in established business processes aimed at raising productivity to a still higher level becomes important. The new product development processes of concurrent engineering and front loading (Khurana & Rosenthal, 1998), for example, require companies to undertake a radical business process review. With regards to business process series ranging from product design to development design, product technology, and production, VIN tools comprise ICT that supports real-time sharing and solution proposals for problems and issues relating to individual areas of specialist work (including development and production) among actors who are physically separated. The combination of CAD (computer-aided design) and VIN tools, especially, demonstrates the roles of sharing common knowledge (Carlile, 2002; Cramton, 2001; Star, 1989) among actors supervising different specialist work and strengthening links among organizations through deep understanding of individual business processes among actors.

Training to enhance the skills of individual actors in order to improve productivity, business processes, and efficiency is also an important theme among many companies. The main focus of ICT-based knowledge management (see Figure 1.5) in the Step process is the sharing of explicit knowl-

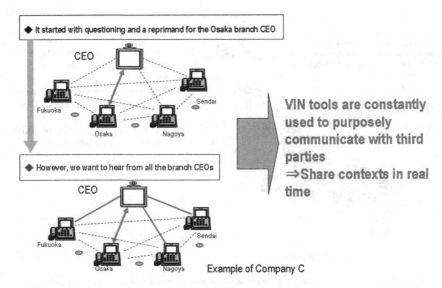

Source: Compiled from interviews with NTT Bizlink and presentation materials for use outside the department.

Figure 1.13.　VIN as a Third Party–Oriented Communication Tool

edge. The transfer of tacit knowledge comprising technology and expertise almost never succeeds. Generally, the technology and expertise (tacit knowledge) of the specialist and the skilled worker can be acquired through the daily practice of on-the-job training and the apprentice system. In recent years, however, ICT and precision machine engineering have progressed to the point where ICT can replace production processes based on the skills and know-how of expert workers. Die production is a typical example of this process. Meanwhile, in specialist domains where it would be difficult for ICT to take over in every area, the only means to learn is for actors to practice the skills and expertise of their seniors at the worksite.

VIN tools can also support efficient and effective training in such cases. At Toyota Motor Corporation, skilled worker processes and expertise are stored on realistic video clips and used to create visual manuals as training materials (Sakai & Amasakai, 2005). These video manuals are one of the concepts for VIN tools with storage functions for video on demand. Toyota's visual manual is also translated into many languages and contributes to training, improving the skills of production line workers in various countries. In this way, Toyota's experiments can be interpreted as converting and storing the tacit knowledge of skills and expertise on a visual database of explicit knowledge. The process of converting tacit knowledge that is difficult to transfer into explicit knowledge through ICT activities is an important theme for advancing knowledge management for high-tech companies.

Matsushita Electric Inc.'s former CEO Kunio Nakamura (2005) noted that there is no management innovation without ICT.

> The significance of IT activities does not end at the production worksite. Every kind of business process must be reviewed. The use of IT enables invisible, tacit knowledge to be converted to visible, explicit knowledge. Tacit knowledge is always present in people's heads. It can't be shared as is, but it can be universalized as explicit knowledge through IT and used to innovate various business processes.

The third category is the promotion of creativity management through inventing new business models and driving innovation. The theme of this third area, especially, is the essence that determines a company's competitive edge. In recent years the integration of rapidly developing heterogeneous technologies, the diversification of business models, and the acceleration of Internet business has speeded up specialization, fragmentation, and labor division in companies and industries. Alongside this, the integration of diverse knowledge dispersed within and outside the company (Kodama, 2007c) and the still more complex integration of boundaries among organizations (internal and external) are becoming urgent themes for companies. Because of these challenges, management leaders have to generate strong organizational capabilities giving birth to creative new technologies and products through deep sharing of information and knowledge as well as rapid decision making.

To drive this kind of creative management, there is a need to do away with divisional partitions and promote mutual understanding among different divisions. The larger the company, the greater the sectionalism in the tasks of marketing, product development and planning, and design and production. The organizational boundaries and the knowledge boundaries of actors' individual specialist fields are areas that create innovation at the same time as creating friction, confrontation, and contradiction (Leonard-Barton, 1995). New ideas and innovation from established research arise easily during dialogs among gatherings of actors with different qualities (see, e.g., Kodama, 2001a, 2002c). It follows that VIN tools take on the role of networking actors from different divisions or specialist fields and sparking interactive communication (boundaries communication). The VIN tools then become the driving force leading to creative management, by which actors innovate new products and business models.

VIN TOOLS AS A DESIGN FOR BA AND COMMUNITIES OF PRACTICE

A large number of management leaders have been under the misconception that VIN tools are a substitute for business trips. In fact, the essential significance and meaning of VIN tools is as a new communication structure. An example is the unique participation styles created among conference members using VIN tools:

> In the past, whenever we had a general meeting, the head office delegation naturally had the lion's share of the discussion, since they attended in large numbers. With videoconferencing, however, the contribution of the factory rose, and it also became easier to express matters that we had previously found difficult to talk about. Perhaps this is because there is often only one seat, depending on the system used. Even if speaking at conventional conferences left us looking stiff, here we were able to talk to each other in a composed manner. (Manager of Factory A)

An important viewpoint is that some remarks should be made by VIN tools even when at the same meeting. For information that cannot be obtained face to face or at a general meeting, it is important to be formed from Ba (Nonaka & Konno, 1998)[1] and communities of practice (Wenger, 1998)[2] that draw out and share VIN tools. All active users of VIN tools recognize that VIN tools are not simply replacements for general gatherings. They also recognize that Ba and communities of practice also occur and are established as new meetings without other ICT tools, such as groupware and email. The opportunity presented by the introduction of VIN tools has launched the richly featured meeting as a Ba or community of practice, which did not exist previously. New data and knowledge flows among actors, who also have the potential to migrate to new behavior. VIN tools can also be used to increase the number of meetings. These can be large meetings with the essential significance of new forms of communication among actors who needed to set up meetings but lacked the means.

Figure 1.14 shows the structure of an in-house meeting at a major sales company using VIN tools. This diagram shows the existence of various Ba and communities of practice for regular and irregular meetings including management meetings at top management level, development and product meetings at middle management level, and store, store operation, and interstore meetings at workplace level.

Actors become able to communicate closely among themselves using the VIN tool environment alone without moving from their fixed physical space. They also become able to set up flexible Ba and communities of practice, and implement new meeting designs. Moreover, actors possessing the essential information and knowledge can be moved to the essential

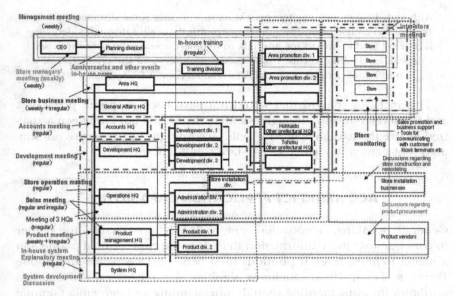

Source: Compiled from interviews with NTT Bizlink and presentation materials for use outside the department.

Figure 1.14. Setting and Design of All Types of Meetings as Ba and Community of Practice (Example of C Company)

location (environments where VIN tools can be used, which can be any-time, anywhere with the use of wireless devices, as mentioned below) and rapidly form Ba and communities of practice. In this way, the dynamic application of VIN tools becomes an enabler for the design of new Ba and communities of practice for actors.

> We are using them among our HQ's business division and the factory's pro-duction division. If we have a customer complaint, for example, the manage-ment at HQ can display the area at fault to the local site during a videoconference. This way we can immediately understand the problem. We never had this kind of conference in the past. A picture is worth a thousand words, as they say. It's also easier for people at the production site to take ini-tiatives if they can get a direct view. (Manager, A factory)

VIN tools are not a system for cutting down on actual gatherings. It is just the opposite—with VIN tools, the number of meetings increases. But rather than increasing the number of unnecessary meetings, meetings take place that, though necessary, could not have been initiated without intro-ducing VIN tools.

VIN tools also make it possible to establish and strengthen the kind of social relationships that would not otherwise occur.

Although I might not pay frequent visits to worksites some distance away in the normal course of business, linking up with these sites through videoconferencing enables me to see and talk to people I would otherwise rarely see or speak to. Thanks to this, I have come to understand the characters of a large number of people. (General Affairs supervisor, A Company)

This includes content that would not have been supposed previously, and social relationships, though essential, that would not have been actualized before the existence of VIN tools. Accordingly, VIN tools are not simply a replacement for actual face-to-face communication.

ACQUIRING ICT APPLICATION CAPABILITIES

Generally speaking, videoconferencing with VIN tools enables the sender to communicate more information than does the conventional phone. On a media richness level, it positions itself between face-to-face and phone (Kydd & Ferry, 1994). Richness signifies the media capabilities and attributes for implementing shared understanding and agreement among interactively communicating actors (Daft & Lengel, 1984). The media with the greatest level of richness is, of course, face-to-face. Would it be possible to say that the VIN tools with the level of richness closest to face-to-face communicate large amounts of information to the receiver, and raise the information and context sharing level among actors, whatever the situation? Among many of the VIN tools applications surveys carried out up to now, it was clear that the media richness level depended greatly on the ability of the actors to apply the tools.

The following points are important in exploiting VIN tools and raising media richness levels. The first is that of the actors' new media and literacy capabilities. The second is the actors' video production capability. All types of communication media, including phones and email, require the knowledge and manners to use that media. These include the tone of voice and verbal ordering in the case of phones, and use of "netiquette," the written word, cc and bcc in the case of email.

In the case of visual communication, however, a different type of literacy is required. VIN tools, to put it another way, require a new type of media literacy. Specifically, as shown in Figure 1.15, the positioning of image corners, background, clothes, facial expression, gestures, camera, and other factors have a great impact on how much the partner understands and how much context is shared. These elements determine the communication quality and context that will affect the partner through the network.

For example, a physical VIN environment of noise pollution and inappropriate gestures and camera settings on the part of the presenter impart

Image = Viewing angle, background, clothes, facial expression gestures, camera

Viewing angle, camera   Background   Facial expression, gestures

Source: Compiled from interviews with NTT Bizlink and presentation materials for use outside the company.

Figure 1.15.   New Media Literacy for VIN Tools

a sense of discomfort and disharmony to the partner, and make it difficult to create an atmosphere conducive to transcending detailed information sharing in order to include important decision making. Actors with low levels of media literacy are unable to share context. Accordingly, a new media literacy capability becomes necessary in order to maximize the richness of VIN.

An important aspect of the second point is that VIN tool users require visual production capabilities. Specifically, the most appropriate media (whether photographs, video, or data) must be chosen for the content to be communicated and used in appropriate combinations to activate VIN tools as visual communication with multimedia functions in the most effective way (see Figure 1.16). These components are added to the media literacy mentioned above to determine the communication quality and content to be communicated to the client. With product presentation and business training, for example, it is necessary not just to exploit the explicit knowledge of documentaries, videos, and products, but to combine with other media in response to meeting contexts and devise methods of showing and expressing these contexts to the partner. This suggests the need to change the usage method for VIN tools through trial and error, depending on the circumstances. It is also essential to effectively communicate to the partner elements that are close to tacit knowledge, such as gestures, body language, and subtle nuances and tensions. If this image producer capability is low, actors are unable to share sufficient context, and their data sharing level may also fall.

In order to acquire these kinds of media literacy and image producer literacy capabilities, actors need to raise their ability to put these into practice through the study process using VIN tools. Generally speaking, the richness of electronic media lies not in the inherent qualities of each medium, but in the opportunity for actors to improve their experienced knowledge

Example: Explanation of videophone connection

* Showing a friendly expression while
  giving explanations

* Explaining connections and functions
  with a reproduction of the real object

* Detailed explanation with data

Choosing and handling the optimal media to support the communication content is essential

Source: Compiled from interviews with NTT Bizlink and presentation materials for use outside the company.

Figure 1.16.　New Application Capabilities for VIN Tools

through "learning by doing" (Carlson & Zumd, 1999). Accordingly, improving the ability to put things into practice through training makes it possible to raise the richness level of the VIN tools and aim to reduce equivocality among actors (Weick, 1979).

THE ENACTMENT OF NEW BUSINESS STRUCTURES THROUGH VIN TOOLS

An example of activating individual VIN tools is given in Figure 1.17. In the future, it is clear that the use of these tools will expand well beyond simple videoconferencing. In addition to interpersonal communication in such areas as education, training, and discussion, they will be active in image transmission, such as monitoring and relays (including products and machines) between people and machines. In the future, further expansion into ubiquitous image networking among machines is also conceivable.

　Looking at matters from an agency perspective (see, e.g., Barley, 1986; Orlikowski, 2000; Orlikowski & Barley, 2001; Robey & Sahay, 1996), actors belonging to different companies and industry types will have different strategic and organizational contexts, so that methods of use differ even with the same VIN tools. The result is that the enacted structures (such as business behavior and forms created as a result of actors' training with regard to VIN application methods and ICT tools) also differ (see, e.g., Orlikowski, 2000).

	All industries Shared	Manufacturing	Distribution	Transport & travel	Services	Broadcasting	Finance	Medical care & welfare	Education	Security	Construction & real estate
Meeting	• In-house meetings • Management meetings • In-house training	• Design meetings • Interworking • Trial product meetings	• Shared knowledge of trial product (convenience stores, etc.)						• School exchanges		
Education Training Guidance		• Technical guidance • Engineering support	• Product explanation • Display guidance • Sales supervisor guidance	• Health check prior to starting work	• Sales methods & service guidance		• Guidance and work support from HQ to branch offices	• Medical training • Operation guidance	• Remote lessons		• Construction site monitoring work
Monitoring		• Monitoring system operation	• Store monitoring • Unmanned facility (store) monitoring		• Store monitoring (eating establishments such as taverns and revolving sushi)	• Broadcast equipment (unmanned) monitoring	• Monitoring of ATMs etc.	• Patients, hospitals, and patient's homes		• Link with crime prevention monitoring system	• Construction site monitoring
Discussion							• Financial negotiations (retail)	• At-home medical care • Health discussion • Link with care support system			• Remote operation of model rooms
Relay			• Perishable goods auction (flowers, food etc.) from market to retailers • Pet shops	• Wedding location hookup	• Intermediary between TV program and actors			• Academic meeting relay			
Other		• Confirming purchases (manufacturing to distribution)	• Direct sales	• Worship (religious organizations)	• Video server connection provides support for sportspeople etc.		• Customer consulting				• Establishing apartments and newly built homes

Source: Compiled from interviews.

Figure 1.17. Usage Structure of VIN Tools

Figure 1.18 shows the usage structure for the VIN tools business field, and Figure 1.19 shows the structure for the medical and welfare fields. Figure 1.20 classifies usage methods around the axes of these VIN tool business (B2B, B2C, and C2C) and communications structures. VIN tools are not limited to the business field (B2B), but are creating consumer-oriented e-business (B2C) and new structures of use (C2C).

Actors are improving the technology functions of VIN tools in response to market environments and social demand, and creating differentiated usage structures and business models with a range of types and business conditions. Put another way, if structuration theory (Giddens, 1984) and agency theory (see, e.g., Emirbayer & Mische, 1998; Jones, 1999) are incorporated, when the definition of human agency is limited to the physical structure of ICT, human agency will transform its own thoughts and actions while interacting with ICT (the physical functions and application methods of the human agency and ITC), and create new structures including usage structure, habits, and business models.

I think that the trigger for introducing and applying VIN tools is the structure of the community (see Figure 1.21). One aspect is the business community structures for the purpose of sharing a range of information and knowledge. These are equivalent, for example, not just to business meetings within the company but to Ba and communities of practice

◎ Discussions in specialist areas	◎ Easy ordering support	◎ In-house meetings
Major banks install in all branches and implement support for the weekends and for specialist fields, such as investment trust discussions.	Requests from customers throughout Japan are confirmed visually on the network and marketed. Also used when developing new products.	Used to connect HQ to branch offices and other sites, implement regular meetings, and facilitate in-house business.

Source: Based on interviews with the author.

Figure 1.18.   Examples of VIN Tools in Business

◎Home treatment	◎Emergency and critical care treatment	◎Support for sign language
Remote medical treatment from home has been launched in regional clinics for patients such as the bedridden or terminally ill, who find it difficult to travel to hospital.	Specialist doctors at emergency and critical care treatment centers support young doctors throughout the country by videophone.	Sign language support from an interpreter after connecting to a sign language support center by videophone

Source: Based on interviews with the author.

Figure 1.19.   Examples of VIN Tools in Medicine and Welfare

including entrepreneurial, industry group, franchise, and NPO and volunteer group communities.

Another aspect is the structure of local authority communities centered on municipalities, or Ba and communities of practice to deliver information flexibly to the public and welfare sectors. A third aspect is Ba and communities of practice as knowledge-based business (Kodama, 1999b, 2000)

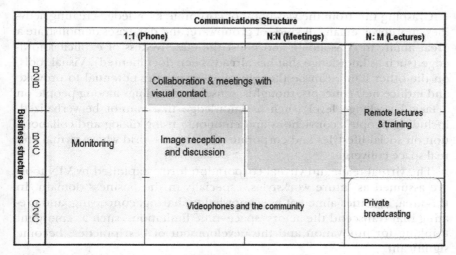

		Communications Structure		
		1:1 (Phone)	N:N (Meetings)	N: M (Lectures)
Business structure	B2B		Collaboration & meetings with visual contact	Remote lectures & training
	B2C	Monitoring	Image reception and discussion	
	C2C		Videophones and the community	Private broadcasting

Figure 1.20. Applicable Domains for VIN Tools

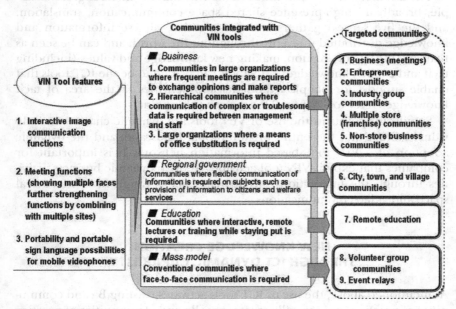

Figure 1.21. Creating Communities with VIN Tools

delivering new value to customers including private-sector companies' own medical, welfare, and education services. I believe that the role of VIN tools is to support community structures as the Ba for application of the above, and to contribute to structures for shared value and trust within the communities.

Grasping this from the viewpoint of corporate knowledge-creating activities, moreover, email, all kinds of groupware, and databases demonstrate a great ability to consolidate and boost the effectiveness of explicit knowledge (such as knowledge that has already been documented). Visual tools, on the other hand, can be thought of as having the potential to provoke and induce new concepts, thoughts, senses, and feelings among people on a tacit knowledge level (such as knowledge that cannot be verbalized, including people's convictions and emotions) using dialog and collaboration on social lifestyles and corporate activity in a world where virtual and real space converge.

The virtual team and virtual corporation forms exploited by VIN tools are assumed as future workstyles, especially in the business domain. In this area, activities aimed at new knowledge sharing, conceiving, and creating that transcend the actors' space–time limitations, such as emergent thinking for innovation and the development of best practice, become significant.

The advanced video communication environment (including, for example, broadband high-presence shared space communication, translation, and the five senses) gathers and ties together diverse information and knowledge distributed unevenly throughout the world, and can be seen as realizing the age of creation, making new knowledge and values (including tacit and explicit knowledge). VIN tools can be said to be the ICT tools that enable support for people's thoughts and activities in the area of tacit knowledge (refer to Chapter 2 for more on this point).

Significant points in the use of VIN tools are dynamic changes in individual culture and the entire organizational landscape, and the dynamic adoption of a VIN tools-based information network. It is important for companies that exploit VIN tool activities to drive dynamic business activities through the interaction with virtual space through face-to-face real space and video communications.

NEW KNOWLEDGE CREATION
THROUGH ICT DYNAMIC CAPABILITY

As mentioned above, the use of ICT tools activates existing Ba and communities of practice, and formally creates new Ba and communities of practice as human networks. New contexts are formed dynamically (transforming existing contexts) through the creation of these networks. These dynamically formed contexts create new meaning with Ba and communities of practice, and go on to create new knowledge aimed at shared strategic aims. The formation of human networks as Ba and the communities of practice goes beyond formal and informal meetings using VIN tools. The

Ba comprising temporary CFTs (cross-functional teams), task teams, and project teams also create new contexts and knowledge through dynamic interaction between real space and virtual space using VIN tools.

Through the use of VIN tools, actors intentionally or unintentionally cross organizational and knowledge boundaries, create new contexts, and form Ba and communities of practice as human networks. The actors' subjugation of organizational and knowledge boundaries, and the contextual architecture capabilities of actors led by new perspectives, create new contexts among actors. Then these new contexts go to create boundaries consolidation capabilities aimed at new human network constructs for actors.

VIN tools complement the actors' capabilities for context architecture and boundaries consolidation. The use of these tools promotes the actors' abilities in these areas and promotes the further use of VIN tools. Then the new perspectives on management leaders' ICT lead to new interpretations and realizations from the dynamic use of ICT. The actors enhance their capabilities for context architecture to create new contexts and boundaries consolidation to build human networks while dynamically interacting with strategic and organizational contexts within and outside the company. Next, the interaction of the actors' capabilities for context architecture and boundaries consolidation further promotes the use of VIN, and enhances actors' ability to use ICT. This ability does not simply enhance the media richness mentioned above. Rather, techniques to raise the level of media richness are one of the means to enhance the capability to use ICT. The essence of the ability to use ICT is productive interpretation and creative realization rooted in the actors' new perspectives (Wisemann, 1988) (see Chapter 2 for further details).

These three elements—capabilities in context architecture, boundaries consolidation, and using ICT application—interact while creating new knowledge from the community knowledge creating cycle (explained in Chapter 2). Then the interaction of these elements creates an organizational capability (called an "ICT dynamic capability" in this text; see Chapter 2 for more details) using ICT, and creates dynamic groups of VIN use, context formation, and human network formation, as shown in Figure 1.22.

FUSION WITH WIRELESS BROADBAND

The typical video terminals used in wireless technology are currently third-generation mobile phones. Transmission speeds for videophones are 64kbps one-way, while video streaming transmits at several hundred kbps. High-speed IP videophones and IP videoconferencing systems have also been realized through wireless LAN technology.

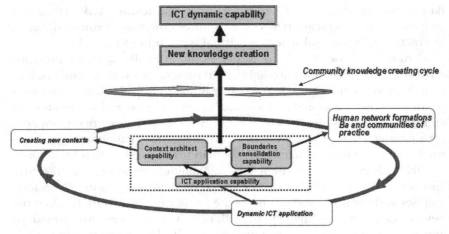

Figure 1.22.   Dynamic Loop for ICT Activity

One appeal of wireless technology is its portability. With no limitations as to location, signals can be sent and received anywhere, at any time. A second reason for its appeal is its temporary nature. It is handy for mobility where fixed-line equipment does not exist. A third reason is its operability. Unlike PCs, mobile videophones, especially, do not require complex operations or setting up with special equipment. Exploiting these benefits, use of remote monitoring and live video relay through mobile videophones are also being actualized (Kodama, 2003b).

One area with great applications potential is linkage with video terminals in the fixed broadband and wireless fields (see Figure 1.23). The "anytime, anyplace, with anyone" high-speed broadband environment is a means of receiving real-time and storable image content (see Figure 1.24). Specifically, it can be seen as a transition from fixed and mobile line markets that function separately to an integrated FMC (fixed mobile convergence) market structure. FMC using VIN tools is already a reality in Japan and France (see Figure 1.24). In Japan, the use of VIN tools through FMC has begun in the corporate and teleworking areas, including the distribution, construction, and maintenance service industries (see Figure 1.25). VIN tools exploiting this kind of fixed broadband and wireless fusion can be formed dynamically anytime, anywhere, through actors' global Ba and communities of practice.

Steve Buckman, CEO of Buckman Laboratories, made the following comment about the impact of wireless technology and video communications on business innovation:

> Broadband connections make online interaction almost as smooth as face-to-face, but a company with any sort of global reach still needs a system that less

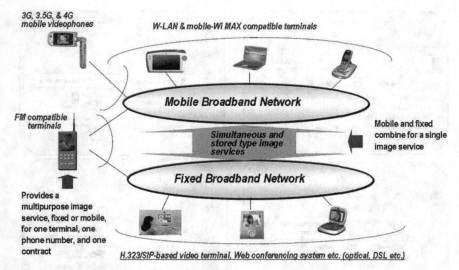

Figure 1.23. FMC Through VIN Tools

Source: Data provided by Leadtek Inc.

Figure 1.24. Broadband Communications between Mobile and Fixed Videophones

connections accelerates and expands, the opportunities for ubiquitous connections anytime and anywhere at high speed will open up radically different opportunities to redefine the equation again. The software that you can con-

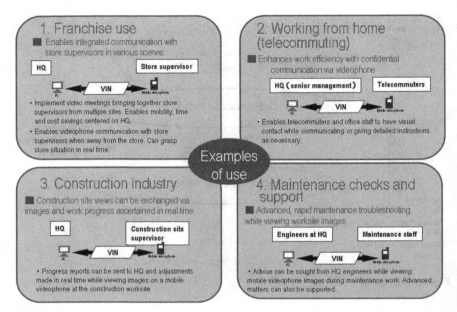

Figure 1.25.   VIN Tool Application Methods Among Broadband Mobile Devices

sider using will have enhanced features that are viable only with high-speed connections. It will allow the merging of data and voice and video into a seamless communication device. This area offers so many opportunities for redefinition that they are too numerous to mention. Focus on wireless and what it can bring to your systems. This will be the foundation for the next big revolution in how we do business. (2003, pp. 241–242)

I think the future will see further development of technology and services, such as the high-speed handover with 3G (and later 3.5G and 4G) phones and wireless LAN. Moreover, fixed landlines and mobile phones have developed amid independently established market structures with separate contracts, terminals, and services. A future environment is being prepared, however, to receive separate fixed and mobile services using a single terminal, phone number, and contract (such services are already in use in Britain and Hong Kong). New services and business models are likely to emerge from the integration of today's fixed landline and mobile communications services into single services.

THE STRUCTURE OF THIS BOOK

The VIN tools platform extends beyond devices for the effective flow of information held by individuals and companies. Regarding information and knowledge in people's lifestyles and corporate activity, VIN tools take on an active role creating new values and social and economic activity in people's corporate lives. VIN tools used for this purpose go to form a range of Ba and communities of practice that should activate social and economic activities with networks that flow with multimedia information extending beyond text to voice and image as well as the knowledge and competence of diverse individuals. Then VIN takes on an important role as an ICT platform that should create new knowledge and values.

The remainder of this text is organized as follows. Chapter 2 considers theoretical frameworks related to dynamic organizational ability (called "ICT dynamic capability" in this text) for companies that have exploited ICT. "ICT dynamic capability" as referred to in this text comprises the three elements of context architecture, boundaries consolidation, and ICT application capabilities, and is discussed from the viewpoint of the new knowledge creation process. The interaction among actors' dynamic use of VIN tools and the community knowledge-creating cycle simultaneously enhances knowledge effectiveness and creativity.

Chapter 3 goes on to consider the decision-making process that introduced and activated VIN tools within the company, aimed at the acquisition of ICT dynamic capability, from the perspective of a strategy-making process. In the ICT-driven community based firm, the merging of the top-down, deliberate strategy-making process and the bottom-up emergent strategy-making process promotes the formation of dynamic Ba and communities of practice inside and outside the company, and simultaneously promotes best practice using ICT and the solving of business problems.

Chapter 4 demonstrates community management frameworks exploiting VIN in the company. The top-down approach (deliberate strategy and implementation) through the innovative leadership of community leaders who comprehensively manage the business community spread within and outside the company (including customers) will enable the dynamic use of ICT by community members. As case studies, I will consider examples of success through the use of VIN tools based on broadband and mobile networks in the finance and automobile industries. In Chapter 5, looking at the creation of business linking industry and academia, I consider cases of VIN development through a bottom-up approach (implementation of emergent strategy) involving members in the strategic community. I make my analysis from the viewpoint of ICT dynamic capability created from dynamic interaction of context and knowledge with customers who have crossed the organizational boundaries among industries.

In Chapter 6, through emergent strategies for small and medium-sized companies, I justify new knowledge in-house after its creation from a trial-and-error process using a bottom-up approach, and I look at the feelings and behavior of actors undertaking to improve in-house productivity and customer services as a strategy for the entire company. In Chapter 7, I take up a Sony case study, and look at an example of VIN tools adoption by means of Sony's deliberate strategy using a top-down approach. Then, learning from the company's mistakes, I consider the process of a successful VIN adoption from a bottom-up approach through emergent strategies centered on worksite organization.

In Chapter 8, I consider successful case studies of VIN adoption by large companies (NTT and Otsuka Shokai). I consider the processes that productively resolve internally generated friction and conflict and transform corporate culture through the use of both top-down and bottom-up approaches. Finally, in Chapter 9, I provide new insights derived from the case studies and theoretical and managerial implications related to new knowledge creation by ICT dynamic capability.

NOTES

1.   Ba is a place offering a shared context. Knowledge needs a context to be created, as it is context-specific. The context defines the participants and the nature of the participation. The context is social, cultural, and even historical, providing a basis for one to interpret information, thus creating meaning and becoming knowledge. Ba is not necessarily just a physical space or even a geographical location or virtual space through ICT (information and communication technology), but a time–space nexus as much as a shared mental space. Any form of new knowledge can be created regardless of the business structure, as Ba transcends formal business structures. Bennett (2001) reported as follows by examining the Japanese characters of "Ba": "The top right character represents the sun; the character on the left the earth; and the bottom right (loosely) rays of light falling to the ground." Thus to the Western (though not the Japanese) eye, Ba is interpretable as a "place of illumination, where sun and earth unite and enlightenment happens."

2.   Community of practice (Wenger, 2000) is rooted in the resonance of value (Kodama, 2001a, 2002c) among the actors. This aspect promotes mutual learning within the community by gaining an understanding of mutual contexts among members and resonating value, and continually generates new knowledge. In the community of practice, the community membership and the community leader at the center of activities are gradually established, and these people dynamically produce the context in which they work toward fulfilling the community's mission.

CHAPTER 2

KNOWLEDGE CREATION PROCESS THROUGH ICT DYNAMIC CAPABILITY

Theoretical Framework

In this chapter I consider the theoretical framework regarding the dynamic organizational capabilities (referred to as "ICT dynamic capability" in this text) of companies exploiting ICT. Many in the top management of the world's leading companies recognize the importance of acquiring a competitive edge through the dynamic use of ICT. I have acquired important knowledge of companies' ICT dynamic capability arising from ICT use and competitiveness. I have gained this from practical experience in ICT marketing, product development, in-house adoption, and customer consulting in the past, and through formal and informal dialogs with corporate ICT developers and promoters of the ICT adoption process. From long-term field studies, it has become clear that ICT dynamic capability arises from the relationship of dynamic interactivity between actors' daily practices and the elements of new perspectives, actors' minds (interpretation and realization), strategic and organizational contexts, and the environmental context surrounding companies.

The actors that emerge in the ICT case studies examined in Chapters 4–8 create new meaning and value in ICT planning, development, adoption, and use. They also create new knowledge by dynamically combining and complementing business activities carried out in real space and those carried out in virtual space through ICT. The following points are important for companies forming this kind of "ICT dynamic capability"[1] with the actors who use ICT.

One point is that actors use ICT to dynamically embed business processes in real space, and link virtual space (ICT applications) and real space (face-to-face) business activities through dynamic interaction. A second lies in the actors' capabilities in forming new contexts and meaning through interaction of real and virtual space aimed at formulating strategies. A third lies in the actors' capabilities in creating new knowledge with high value added based on new contexts and meaning. Specifically, actors possess the capabilities to bridge organizational and knowledge boundaries (Kodama, 2007a, 2007b) and accomplish problem solving and new challenges through dynamic interaction between real and virtual space.

In Chapter 1, I used the term "ICT" in preference to "IT." "ICT" comprehensively includes the meanings of "IT" (conversely, some assert that "IT" incorporates "ICT"), and I have used "ICT" throughout the book. I have also used the term "ICT" when referring to existing research that frequently uses the term "IT." Neither the technical functions and capabilities of groupware, database systems, and ERP (enterprise resource planning) information systems, nor the technical debate on the effects of adoption, are central to this book. The main thrust of my research is to consider in detail the use of telecommunications networks that link individuals through Internet, broadband, and mobile phone, and the relationship between a company's ICT strategy and its organizational capability.

COMPETITIVE ADVANTAGE THROUGH ICT: WHAT IS ICT CAPABILITY?

In this chapter, I want to look back on existing research regarding the organizational capability of companies using ICT, and present the results and practical significance of new research topics. Recent studies have reported the practical business insight that ICT has already become commoditized, and cannot be the source of a company's competitive excellence (Carr 2003, 2004). In the field of academic research, meanwhile, from the strategy research foci of positioning-based and resource-based views, scholars have indicated that while ICT has the potential to raise operational efficiency by itself, the fact that it is easy to copy means that ICT alone cannot become a source of competitive excellence, and is unlikely to be a means of differentiation from rivals. It cannot, therefore, sustain competitive excellence (see, e.g., Barney, Wright, & Ketchen, 2001; Clemons & Row, 1991; Mata, Fuerst, & Barney, 1995; Porter, 1991, 2001).

Moreover, scholars researching the relationship between organizational capability and firm performance through ICT point out the following about gaining a high-value, copy-resistant ICT capability. One aspect is that ICT capability can be acquired through interaction among the three resources of physical ICT infrastructure (hardware and software), human resources possessing ICT skills, and intangible assets enabling ICT (Bharadwai, 2000). A second aspect is the need to acquire ICT capability for trust, coordination, and negotiation competences through communication and collaboration with in-house ICT and business unit managers (Ross et al., 1996).

A third aspect points out that the relationship between an organization and ICT, and the effects of integration among human resources, ICT resources, and business resources involving redesign of the business process can give rise to the possibility of creating ICT capability with a sustained competitive edge (Powell & Dent-Micallef, 1997). A fourth aspect suggests that organizational learning (see, e.g., Huber, 1991) can be grasped as a process developed by companies' new knowledge and competences, and that ICT capability to enhance company performance is enabled through the experience-based study of organizational learning processes within a company (Tippens & Sohi, 2003).

In this way, acquiring and sustaining ICT capability can be thought of as not simply a matter of ICT infrastructure itself and the resources of superb ICT managers. Rather, it enables these resources to effectively interact and cooperate with other resources (such as people and organizations) and company-held contexts (such as changing environment, the strategy formulation and implementation process, and corporate culture) inside and outside the company, thus enhancing company performance.

Generally speaking, the strategic context (strategy formulation and implementation) and organizational context (including the business processes, the decision-making process, organizational structure, in-house political power, and corporate culture) differ from company to company. This creates the possibility of copy-resistant organizational capability and sustained competitive edge differentiating a company from its rivals (see, e.g., Clemons, 1991a, 1991b; Clemons & Row, 1991; Weill & Broadbent, 1998; Weill & Ross, 2004) arising from the embedding of ICT in a company, leading to the mutual benefits of ICT interacting with a company's inherent strategic and organizational contexts.

Meanwhile, systematic research into ICT and productivity by Brynjolfsson and Hitt (1995, 1996) indicated that supplementary investment in intangible assets, such as organizational resources within a company, and business process innovation must take place simultaneously in order to enhance productivity through ICT. The research also suggested that ICT investment enables companies to demonstrate the effects of productivity by aiming to link it with investments in personal and organizational intangible assets. It became clear that the same sum of ICT investment can bring about high productivity gains in one company and low gains in another (Brynjolfsson, 2000).

Accordingly, from the above debate, it can be said that linking the intangible assets of human resources and human competency with ICT-related resources raises the quality of ICT capability, and contributes to improving company performance and competitive advantage. The research mentioned above presents the new implications, on the academic side, as part of the practical business insight (Carr, 2003, 2004) that ICT cannot become the source of a company's competitive excellence.

Much of the past academic research on ICT capability is content statistically analyzed by large-scale sampling of the connection between ICT capability and company performance. This content centers on the static experimental study of analyzing the structural elements of ICT capability under a fixed competitive environment. Much of the existing research does not ascertain the formulation and implementation of ever-changing ICT strategies in a dynamic, competitive environment on a real-time axis, nor does it analyze ICT capability. In other words, this research lacks dynamic analysis.

Meanwhile, a large number of talented businesspeople face a range of problems and issues on a daily basis. They ask such questions as "How can we structure ICT capability to create a competitive edge?" "How can we reconstruct existing business processes, and use ICT to raise operational efficiency and productivity?" and "How come other companies have had success with adopting this ICT system, and we haven't?" The viewpoint of strategic management as practice (Kodama, 2007b) implementing the "who," "what," "why," "when," "to whom," and "how" approach with regard to these many issues and ICT strategies is important for the concerns of many businesspeople. For this reason, I believe that the construction of dynamic theory (called "ICT dynamic capability" in this book), relating to ICT capability embracing planning, development, installation, operation, and results, will become an important theme of future academic research. The pursuit of this "ICT dynamic capability" may have beneficial managerial implications for businesspeople.

The approaches aimed at building dynamic theory as it relates to ICT capability can be considered from an organizational learning view or a knowledge-based view. In this chapter I consider the process by which a company dynamically implements the organizational learning process of planning, adopting, and using ICT, and arrives at new knowledge or innovation. This chapter also pursues in detail a company's process of knowledge creation using ICT. It is thought that actors inside and outside an organization can embed the temporal elements of dynamically creating new knowledge through the organizational learning process using ICT within the theory of ICT dynamic capability.

In the following chapter, I consider the knowledge creation process for real and virtual space. Then I discuss the ICT dynamic capability of actors that create new knowledge using ICT.

NEW KNOWLEDGE CREATION THROUGH ICT: THEORETICAL FRAMEWORK

In this section, I consider the theoretical framework of knowledge creation activities from actors with regard to real and virtual space (ICT applications). Actors' knowledge-creating activities through daily practice become an important process (see, e.g., Kogut & Zander, 1992; Leonard-Barton, 1995; Nonaka & Takeuchi, 1995) for acquiring the intangible expertise and skills of tacit knowledge (Polanyi, 1966) and practicing knowledge (Schon, 1987; Suchman, 1987).

This kind of tacit and practicing knowledge does not only arise from the activities of actors in real space. I think it enables the inspiration of new tacit knowledge and the interaction of tacit and explicit knowledge with regards to virtual space through the dynamic activities of high-level ICT (such as the VIN tools presented in Chapter 1).

Knowledge management crossing real and virtual space, which is currently practiced by a large number of companies, comprises activities that create high-quality tacit and explicit knowledge among actors and enhance a company's ability to compete and grow. One important focal point for new value creation and innovation is the deliberate formation of diverse communities of practice and Ba in real and virtual space, both inside and outside the company, by individual employees including customers and partners. Another is the dynamism that constantly creates new knowledge through a spiral transformation process of communities of practice and Ba as it relates to tacit and explicit knowledge.

First, I outline the significance and essence of knowledge management. Then I discuss the knowledge creation process in real and virtual space under an ICT environment.

The Significance and Essence of Knowledge Management for the ICT Era

Corporate knowledge management in an ICT environment has two meanings. First is the pursuit of efficiency and productivity in corporate activities. Specifically, this involves areas such as enhancing the daily business process and transferring and sharing best practice. These areas include the sharing of existing knowledge assets where actors have demonstrated ICT dynamic capability, and a range of activities to improve management quality. Second is the pursuit of creativity in corporate activities. This involves grasping knowledge as a source of an organization's competitiveness and managing creatively with innovation transcending best practice. Put another way, it involves a company creating (producing) knowledge essential for acquiring sustainable competitiveness, and the strategies and processes to share and exploit this knowledge.

Specifically, it doesn't just involve actors demonstrating ICT dynamic capability while enhancing the efficiency and productivity of business activities through systematically consolidating and sharing individual knowledge and corporate knowledge assets. It also involves sustainably creating value and innovation in new products and services as the fruits of new knowledge. Accordingly, ICT dynamic capability is also organizational capability inherent to the company whereby actors pursue both efficiency and creativity while maintaining a competitive edge.

At present, for many companies, "knowledge" is the most important source of permanent competitive advantage (Leonard-Barton, 1995; Nonaka & Takeuchi, 1995). Today's companies are facing numerous issues including dramatically changing markets, diversifying and advancing technology, corporate competition crossing industries and adding layers, and products with shorter lifecycles that rapidly lose their novelty. Companies that thrive under these conditions are able to constantly create new knowledge, distribute it widely through an organization, and speedily incorporate new technologies and products. Put another way, a company's greatest mission is to innovate endlessly.

Accordingly, whether focusing on real or virtual space, it will become increasingly important to take communities and Ba for new knowledge creation activity through interaction of an organization's individuals, concentrate ICT dynamic capability with diligent daily habits, and sustainably create new knowledge. What is required of the innovative company in the 21st century is to go beyond cutting transaction costs and implementing operations efficiently through ICT activities to the endless creation of new knowledge as a source of corporate competitiveness as the essence of knowledge management.

Knowledge Creation Process Through the SECI Model

Two kinds of knowledge exist: tacit and explicit. Tacit knowledge is individual, subjective knowledge that is difficult to put into words or communicate in other ways. Tacit knowledge corresponds to conviction, image, intuition, skill, and expertise. During their daily activities, however, people need to understand matters as individuals and communicate meaning to others. This requires the capacity to verbalize and objectify a range of tacit knowledge. This corresponds to another type of knowledge: explicit knowledge. Explicit knowledge takes the shape of formalized documents such as books and manuals. The main point of sharing explicit knowledge is, basically, to compare the tacit knowledge in people's heads and the routine activities of embedded behavior, and share information and knowledge using ICT.

So both these types of knowledge are important. Since tacit knowledge is a major source of explicit knowledge, the creation of new knowledge from such areas as new products and services is essential. Tacit and explicit knowledge are not completely independent, however, and at innovative companies they circulate with endless dynamic interaction, mutually benefiting each other (Nonaka & Takeuchi, 1995).

Knowledge creation is the process of recreating knowledge acquired from a corporate environment on each of the personal, group, and whole-company levels. Knowledge creation is a mutually transforming process through an interactive spiral of tacit and explicit knowledge. Next, I discuss four modes of knowledge transformation (see Figure 2.1).

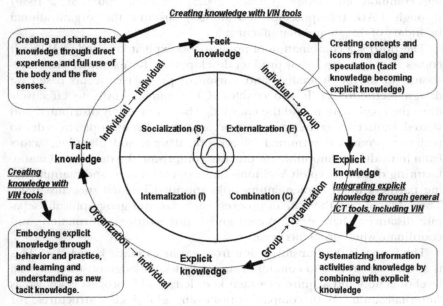

Source: Nonaka and Takeuchi (1995), modified

Figure 2.1. Process of SECI Model

The first is the process of transforming individual knowledge into organizational knowledge. This is called socialization. It is a process that exploits the body's five senses, and shares and creates tacit knowledge through direct experience. This is equivalent, for example, to someone learning a craft through daily practice and on-the-job training at the worksite. There is also the case of socialization using VIN tools, as explained in Chapter 1. One example is creating a place-of-sale network, and sharing and experiencing the tacit knowledge of instructors and experts in virtual space through management training in real time. The case of Toyota Motor Corporation's visual manual, mentioned in Chapter 1, involved putting the tacit knowledge of experts and workers on video, sharing the content on demand through the network, and embedding the workers' skills and expertise through their own images.

The second is externalization as a process that transforms tacit knowledge to explicit knowledge in order to express the tacit knowledge in clear language and concepts. Through this process, actors talk and speculate amongst themselves to create concepts, language, and designs, and the individually accumulated tacit knowledge becomes explicit knowledge to be shared in groups. Examples of externalization using VIN tools are the creation of explicit knowledge from tacit knowledge through sharing images and concepts via VIN tools, and the actualizing of images by sharing common knowledge (Carlile, 2002; Cramton, 2001; Star, 1989) through CAD (computer-assisted design) around the organizational boundary of design and manufacturing.

The third is a combination of implicit and explicit knowledge. When it comes to a company's new product development, for example, this corresponds to thorough analysis and pursuit of product concepts based on design methodology. It also enables ICT dynamic activity to effectively drive the combination, and the knowledge becomes widely distributed and shared, transcending the group and organization. For example, in order to realize a product determined by concept, design, and planning, actors from individual communities of practice comprising the design and manufacturing divisions exploit VIN tools and external design and manufacturing partners when implementing tasks specified in each specialist field. They also confirm the state of progress each day over geographically separate distances, share development and manufacturing information, and coordinate when problems occur.

The fourth is the transformation from explicit to tacit knowledge. This is a learning process to communicate explicit knowledge on the practical level of action and acquire new tacit knowledge. This process is known as "internalization." With companies possessing global sales structures, for example, internalization involves sharing customer sales information such as daily proceeds and customer feedback in real time using VIN tools and, further, actors reflecting among themselves. Actors then follow up by swiftly implementing specific actions in response to market needs.

Above, the process of communicating explicit knowledge on a practical, action level through the process of socialization, externalization, combination, and internalization (the SECI model) and inspiring new tacit knowledge comprises the entire process of knowledge creation. The core concept of the knowledge creation process is the seamless process of dynamic interaction between tacit and explicit knowledge. Through this sustainable spiral SECI model, the organization acquires new knowledge and achieves innovation.

Empirical Background of the SECI Model

Next, I consider the relationship between the SECI model and ICT (VIN tools) based on empirical background. The first aspect I look at is the development of the "i-mode" service, one of the world's leading services for Internet connection using mobile phones. The realization of the i-mode service is a good example of new knowledge created as a result of the SECI model's knowledge creation process (Peltokorpi, Nonaka, & Kodama, 2007). For this series of business processes ranging from planning and developing i-mode to introducing services, the main business seen was, of course, knowledge creation activities as the strategic actions of actors in real space. But it was ICT activities that demonstrated the power of that information and context-sharing process among actors in the distributed geographical environment of the nationwide offices of major company NTT DoCoMo, and later on, the worldwide partnership aimed at the global development of the i-mode service.

With the process of socialization and externalization, for example, development project members share information and contexts. They also spark the creation of new knowledge using VIN tools among actors inside and outside the company (including development partners and customers at home and abroad) aiming to establish the i-mode service concept. The combination and internalization process, moreover, uses VIN tools with partner companies and relevant divisions inside and outside the company, and promotes close links while aiming to share information and context with regards to the realization of the service (establishing service development, system design, system construction, management systems, support systems, and other individual business processes) and feedback from the customer and markets after launch of the service. For i-mode development and the realization of global service development, knowledge sharing and inspiring using VIN tools with actors distributed throughout Japan and the world was an important means of positioning.

As mentioned in Chapter 1, VIN tools are a network strategy tool with a still stronger affinity for a management innovation/customer value creation model than email and groupware. Perceived from the viewpoint of knowledge creation activities, ICT tools, such as email and all kinds of groupware systems and databases, have demonstrated great power as a combination step creating a transformation stage from implicit to explicit knowledge with the SECI model (Kodama, 1999b).

Meanwhile, VIN tools enable the induction of new concepts, feelings, and senses at the level of tacit knowledge through dialog and collaboration on social lifestyles and corporate activities in a world where virtual and real space converge. The telemedicine and e-learning mentioned in Chapter 1 are typical examples of this. Moreover, actors effectively and efficiently enable the demonstration of ICT dynamic capability embedding VIN tools into the business processes (marketing, development, manufacturing, sales, distribution, and support) of integrated supply chain management. The more advanced video communications environment of broadband enables the creation of new knowledge and values by gathering, analyzing, and linking all kinds of information and knowledge that is distributed unevenly throughout the world. VIN tools are the ICT tools that enable support for actors' thinking and activities with the SECI model.

Knowledge Creation Model Through Knowledge-Handling Process

In the previous section, I mentioned the relationship between ICT tools and the process of knowledge creation from the interaction (SECI model) of tacit and explicit knowledge. To begin with, the knowledge that is practically necessary for an actor's business activities cannot clearly be separated into tacit and explicit knowledge grasped upon a specific temporal axis (Tsoukas, 1996). Actors' thoughts and actions are always embedded in mind and body on a path-dependent, tacit knowledge base, and actors embed in real time a range of tacit and explicit knowledge from a company's dynamically changing internal and external environment. From this it follows that the important focal point is not to separate and debate categories of knowledge, but for actors to explain the knowledge creation process individually and among each other with an embedded knowledge handling model.[2]

Previously posited models described organizational learning (Huber, 1991; Walsh & Urgson, 1991) and the innovation process (Damanpour, 1991) as knowledge-handling models. The knowledge-handling process involves types of processes (see, e.g., Carlile & Rebentisch, 2003; Hargadon & Sutton, 1997; Kodama, 2000, 2002a, 2002b) dealing with knowledge embedded in actors and organizations including knowledge transfer (see, e.g., Argote, 1999; Szulanski, 2000), knowledge access (Grant & Baden-Fuller, 2004), knowledge acquisition, knowledge storage, knowledge retrieval (Hargadon & Sutton, 1997), and knowledge transformation (Carlile & Robentisch, 2003).

As previously explained with the SECI model, this knowledge-handling process is also logically and practically useful as a model whereby actors within and among organizations explain the knowledge creation process.

Research conducted in the past, however, emphasized the knowledge-handling process in the real space of the workplace, and there were few research findings on the knowledge-handling process in the workplace environment comprising virtual space, or a blend of real and virtual space, through ICT. The purpose of this book is to clarify ICT dynamic competences as actors and organizations through the knowledge-handling process exploiting ICT in an increasingly distributed broadband environment.

Considering the knowledge-handling process as it relates to virtual and real space, it became clear that axis of the handling process, which is at the core of the knowledge creation process from the viewpoint of the writer's previous practical experience and field surveys, comprised the elements of knowledge accumulation, knowledge sharing, knowledge inspiration, and knowledge creation (including the meaning of knowledge integration and knowledge transformation) (see, e.g., Kodama, 2000, 2002a, 2002b, 2004).

Actors create new knowledge through the formation of business communities (specifically, these are new products, services, and business models, which I have decided to refer to as "community knowledge" in this book), and apply them in the workplace. Then knowledge comprising competence, skill, and expertise is embedded and accumulates within actors themselves and among actors (organization). In this book, I refer to this series of knowledge-handling processes as the "community knowledge creating cycle," a concept that I now explain further.

Community Knowledge Creating Cycle in Real and Virtual Space: Theoretical Framework

The community knowledge creating cycle (see Figure 2.2) comprises the four processes of knowledge sharing, knowledge inspiration, knowledge creation, and knowledge accumulation.

The process of sharing involves ample dialog resulting in understanding between concerned parties regarding the vision and objectives pursued by different organizations seeking to understand and share each other's knowledge. The process of inspiration through contact involves inspiring and multiplying various aspects of community knowledge within the organization to identify

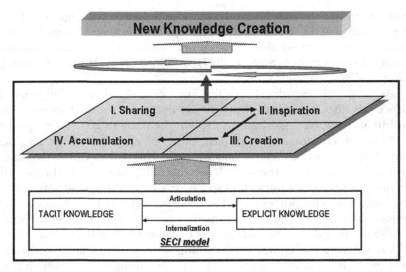

I.   Sharing   Understanding and sharing existing knowledge held in the community
II. Inspiration   Propagating knowledge through inspiration related to existing knowledge
III. Creation   Creating new knowledge
IV. Accumulation   Storing diverse new knowledge born from the process of inspiration, propagation, and creation,
                    including know-how accumulated through the practical application of community knowledge in the market

Source: Kodama (2006)

Figure 2.2.   New Knowledge Creation through Community Knowledge Creating Cycle

problems, challenges, and solutions so that visions and concepts can be realized on the basis of community knowledge shared by the different organizations concerned. The process of creation involves creating new community knowledge on the basis of community knowledge inspired and multiplied within the circle of organizations concerned. The process of accumulation involves methodically (as an organizational effort) accumulating within the community, as valuable expertise, the various aspects of the community knowledge garnered through the processes of sharing, inspiring, and creating.

These processes possess different characteristics in the various business communities operating inside and outside the organization. Various forms of business communities exist amidst inside and outside the company environment: relationships within the same company, among companies, and between companies and customers. These can be generally classified into two cases. The first is business communities formed among actors possessing knowledge in the same specialist areas or who have little knowledge difference (Hinds, 1999; Schank & Abelson, 1977). The second is business communities formed among actors possessing different qualities of knowledge.

The former case is called a community of practice, and the latter is a business community that the writer calls a strategic community (see, e.g.,

Kodama, 2000, 2007a, 2007b). First, looking at the relationship between communities of practice and strategic communities (SCs), I would like to consider the actors' communication and behavior with regard to the knowledge boundaries between them.

Three Layers in Knowledge Boundaries

Figure 2.3 shows the relationship between the communication structures of actors possessing different knowledge and the business community structures inside and outside the organization. Actors perceive various organizational and knowledge boundaries amid their daily business activities. The organizational boundaries that I refer to here correspond to boundaries among work duties for formal organizations, such as research, development, production, and sales; boundaries among management layers inside the company; and boundaries among customers and external partners. The boundaries are formed by actors with different backgrounds and specialist areas of knowledge. Put another way, this is the actualization of knowledge boundaries among actors that are always defined. The features of these boundaries are formed from three general layers in response to the degree of novelty and uncertainty among actors, as shown in Figure 2.3 (Carlile, 2002, 2004; Jantsch, 1980; Shannon & Weaver, 1949).

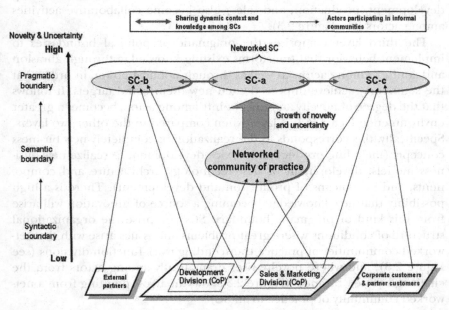

Figure 2.3.   Community of Practice and Strategic Community (SC)

The first layer comprises syntactic and information-processing boundaries to implement the accurate transfer of information and knowledge among actors. In this layer, the level of novelty and uncertainty among actors is low. The layer includes commercialization resulting from routines and established development and production methods as already determined business processes. Observance of in-house syntactic boundaries and the rules of in-house procedures with the syntactic boundaries within the company are emphasized, and the more efficient and productive business processes become the objectives.

The second layer comprises semantic and interpretive boundaries to cultivate new meaning among actors and act to translate new knowledge. The level of novelty and uncertainty among actors rises in comparison to the syntactic boundary. Specifically, actors' incremental actions to further improve existing business processes and development and production methods, and to upgrade products on the basis of path-dependent cumulative technology, occur within the semantic boundaries. In-house semantic boundaries naturally emphasize the observance of syntactic boundaries and the rules of in-house procedures, but these semantic boundaries also drive sets of organizational learning, such as TQM aimed at promoting best practice within the company, and improving business practice are also driven by these semantic boundaries. Incremental improvements and enhanced work practices through networked communities of practice within the company transcending each work function division (including development, production, and sales) also become collaborative activities among actors (see Figure 2.3).

The third layer comprises the pragmatic or political boundaries to implement behavior that transforms existing knowledge through abrasion and conflict among actors as well as through political power, in order that the actors may achieve unprecedented new themes and targets. It follows that the degree of novelty and uncertainty among actors becomes a greater environmental factor in this layer when compared to the other two layers. Specifically, this corresponds to the realization of completely new business concepts (including product and service development to realize new business models, development of new technology architecture and components, and new means of production and development). There is a high possibility that new knowledge becoming a source of innovation will arise from this kind of pragmatic boundary. SCs comprise the organizational structure of conditions where great problems and issues arise with the networked communities of practice transcending work-function divisions (see Figure 2.3), and where novelty rises dramatically among actors from the challenge of new business (Figure 2.3 shows an arrow pointing from a networked community of practice to an SC).

The boundaries that form these three layers, however, are mutually interdependent, and the boundary features change greatly according to environmental change (including customer needs and competitive environments) and actors' intentions and interests (syntactic boundary -> semantic boundary -> pragmatic boundary, and the converse process). In cases where the changes in environment and actors' intentions work more strongly toward innovation and corporate change, the relationship between actors shifts closer to the pragmatic boundary (syntactic boundary -> semantic boundary -> pragmatic boundary).

Previously, ICT tools were constricted by the focus on the technological aspect of how to communicate content accurately and effectively. Considered from the viewpoint of the communication structure mentioned above, the first layer corresponds to the syntactic boundary. As mentioned in Chapter 1, however, the bidirectional functions of communication have become fuller against a background of development in broadband and video transmission technology in recent years, and the multimedia data of information, voice, and images is capable of closely sharing a range of content among actors with a real-time and on-demand structure.

Accordingly, VIN tools promote knowledge sharing and mutual interaction among actors, and not only enable sharing of meaningful issues with regards to semantic boundaries but also enable efficient and effective support for actors' specific decision making and behavior in an environment of pragmatic boundaries with a high degree of novelty and uncertainty. In the next section, I explain the community knowledge creating cycle with regards to communities of practice and SCs.

The Community Knowledge Creating Cycle Resulting from Communities of Practice

Actors within general companies belong to various sections and are in charge of fixed specialist domains. In communities of practice for the same work functions (such as individual specialist areas including sales, development, and manufacturing) (Brown & Duguid, 1991; Lave & Wenger, 1990; Wenger, 2000), the activities of the daily routine are central. Here actors' path-dependent knowledge (March, 1972) is important, and organizational learning activities (Huber, 1991) are implemented daily on the basis of knowledge as organizational memory (Walsh & Urgson, 1991) accumulated (see the accumulation stage in Figure 2.2) by actors in the past. By improving daily work practices in the community of practice, the improvement-related themes (the path-dependent knowledge assessment of the past) in a shared, similar thought world (Dougherty, 1992) are shared among actors through dialog (see the sharing stage in Figure 2.2).

For communities of practice whose main duties are routine, with a low degree of novelty and uncertainty, the difference in domain-specific knowledge among individual actors in specialist areas is relatively low (Carlile, 2004). Common language is activated and problem points and issues related to work improvement are shared within the range of knowledge from these almost identical specialist fields. Similar cases with low degrees of novelty and uncertainty (equivalent to the semantic boundary above) are even found among networked communities of practice (see Figure 2.3) formed from among the different communities of practice of the development, manufacturing, and sales divisions.

Next, issues aimed at solutions to everyday themes are probed, new meaning is created aimed at realizing the themes in the workplace, and this new meaning is shared among actors (knowledge inspiration stage). With this knowledge sharing and inspiring process, actors also use VIN tools to promote frequent knowledge sharing and inspiring at geographically separate workplaces. At the knowledge sharing and inspiring stage, actors are not conscious of whether the space is real or virtual.

New knowledge is created based on sparing, inspiring knowledge among actors (see Figure 2.2's knowledge creation stage). The networked communities of practice at the design and production divisions combining VIN tools and CAD (computer-assisted design) include good examples for describing the knowledge creation stage using ICT. CAD becomes an effective boundary object (Star, 1989) for creating and sharing new meaning vis-à-vis themes and problem points faced among actors. This created new knowledge is then accumulated (the knowledge accumulation stage in Figure 2.2) as the organizational memory (March & Simon, 1958; Walsh & Urgson, 1991) of skill and expertise in individuals and among actors.

In this knowledge accumulation stage, knowledge that can be transformed into explicit knowledge is, of course, accumulated in such locations as the VIN tool's video or document database. In this kind of community of practice (including the networked communities of practice), the tasks of incremental innovation (such as new product development by improving daily routines and upgrading path-dependent resources) are implemented through the organizational learning process of knowledge sharing, inspiring, creating, and accumulating (including ICT activity) mainly on the basis of existing path-dependent knowledge.

The Community Knowledge Creating Cycle Resulting from Strategic Communities

Meanwhile, in cases of new product development and constructing new business models, where the degree of novelty and uncertainty is great, the dependency rises among actors in each community of practice formed as a division of work function (Carlile, 2002). Moreover, for project-based organizations (Kodama, 2007c) comprising projects (Hobday, 1998, 2000) temporarily formed by actors from different backgrounds and various specializations or comprising formal organizations promoting exploratory activities, dependency rises along with novelty and uncertainty among actors, as it does with the cases of challenging new issues that are difficult to solve simply with the path-dependent knowledge accumulated in the past (this corresponds to the pragmatic boundary in Figure 2.3). In this kind of situation, actors frequently fall into the trap of avoiding consideration of new knowledge outside the path-dependent knowledge that they have built up themselves over long years. There is a great likelihood that excellent actors who have had experience of successful models will be constrained by the "curse of knowledge" (Camerer, Lowenstein, & Weber, 1989).

In contrast to the communities of practice where dependency and knowledge difference among actors is small, I have decided to use the term "strategic community" (SC; Kodama, 2007a, 2007b) for the organizational structures that face (or that spontaneously takes on the challenge of high-novelty themes) themes and issues with high novelty and uncertainty. Actors within SCs exploit path-dependent knowledge while dynamically incorporating new and different (path-breakthrough) knowledge aimed at new challenges, and endeavoring to avoid competency traps (Levitt & March, 1988; Martins & Kambil, 1999) and core rigidities (Leonard-Barton, 1995).

The concept of SC is that actors aiming at unknown challenges dynamically access domain-specific knowledge and resources that are deliberately chosen to be different to those available inside or outside the organization (or company) to which they belong. So doing, the actors can go on to form heterogeneous business communities, embedding these knowledge and resource areas as new human networks. As shown in Figure 2.3, these business communities correspond to the in-house SCs (SC-a) transcending divisions (examples include strategy teams to take on mission items and project teams for emergency business strategies), SCs with strategic partners (SC-b), and SCs with specific customers (SC-c).

The SCs that transcend these kinds of cross-functional projects and diverse communities of practice (corresponding to SC-a) and diverse, multiple SCs separated in time and space (corresponding to SC-b and SC-c) form interactively connected, layered, networked SCs (networks linking SC-a, SC-b, and SC-c in Figure 2.3). The phenomena of intense dialog, tension, abrasion, conflict, and discord are observed among the actors (see, e.g., Kodama, 2001a, 2005a).

Within the SCs or networked SCs, while one strategic aim is shared, the individual information, context, and knowledge (tacit and explicit) held by actors is dynamically shared through intense dialog. This corresponds to the sharing stage in Figure 2.3.

This dynamically shared information, context, and knowledge creates further new meaning within an SC. Then the actors work to create action plans for the various problems and issues through means of interactive learning aimed at embodying strategic objectives. In the sharing stage, considering the characteristics of the knowledge boundaries among actors, the translation process (with the characteristics of a semantic boundary) is promoted to generate new meaning among actors (Carlile, 2002, 2004).

Within SCs, however, the debate frequently grows intense due to the clash of intense opinions from the viewpoint of individual actors' own values and specializations regarding specific problem points and themes. The greater the hurdles of strategy aims (such as the degree of difficulty of product development or the complexity of business models), the greater the novelty, and the more uncertain the market, the greater the collisions, abrasion, and conflict generated among actors (Carlile, 2002, 2004). These become cases (or "battles") where actors with individual viewpoints clash with each other, never compromising. There is a high chance that high-quality knowledge creation and innovation will be frequently created from these "battles" in business fields, especially from heterogeneous members (Johansson, 2004; Leonard-Barton, 1995). Many cases of innovative product development are SCs comprising heterogeneous actors always creating new community knowledge (Kodama, 2007b).

Capable actors, however, can wring out policies where community members work together though constructive and creative debate to fully deliver solutions. Conversely, it is essential for SCs to promote contrary propositions synthesized through dialectic dialog (Kodama, 2005b). This is the inspiration stage of Figure 2.2. Meanwhile, in the inspiration stage, looked at from the features of the knowledge boundary, the transformation process (special qualities of pragmatic boundaries) of existing knowledge shared among actors is promoted (Carlile, 2002, 2004). Conflict is resolved productively, and opposing propositions and themes resolved dialectically, through repeated constructive debate regarding activities not just in real space but also in virtual space exploiting VIN tools.

Promoting dialectical dialog at the inspiration stage connects conflict with collaboration. In the creation stage of Figure 2.2, actors collaborate to solve specific problems and themes. They also integrate knowledge distributed within SCs or networked SCs and create new community knowledge aimed at knowledge creation for strategic objectives. New product development, for example, is a process that completes basic and detailed designs such as whole-system architecture and subsystems, and then links them to the completion of product development for specific prototypes and integrated work divisions, including the research, development and design, and manufacturing divisions. Moreover, it is the stage where actors apply and launch to market new products and services by forging strong links among the PR and advertising, sales, distribution, and service divisions. This creation stage has the features of a transformation process reaching how actual products and services are realized, strategy objectives implemented, and markets and customers changed.

Moreover, in the Accumulation stage of Figure 2.2, such actors' competence, skills, expertise, and other assets acquired at the creation stage as well as the community knowledge acquired through business practice (introducing new products and services to the market) in the market and the field accumulates as the resources of individual actors and organizations. At this accumulation stage, actors form SCs as translation processes, self-reflect (Shone, 1983) through implementing strategies to interact with markets (customers), and discover new, secondary meanings through "sense-making" (Daft & Weick, 1984; Weick, 1995). Then actors form new cognitive frames and pursue strategies aimed at the next challenge.

Strategy outcomes are obtained as a result of the formulation and implementation of strategies as series of these community knowledge cycles. These community knowledge creating cycles, however, do not end after a single revolution. They spiral upwards on a time axis from past to present and future (see Figure 2.2). At the accumulation stage, actors return to the challenge of forming knowledge with new pragmatic boundaries from semantic boundaries through a new "sense-making" learning process.

ICT DYNAMIC CAPABILITY:
THE THEORETICAL FRAMEWORK

The above-mentioned SECI model and community creating knowledge cycles share three major features. The first is the actors' focus on acquiring ICT application capability through the practice of organizational learning. The second is the focus on dynamically creating new context and meaning through actors' interactivity with ICT. The third is the focus on actors' transcending, consolidating, and integrating (knowledge integration) an organization's boundaries of differentiated knowledge distributed inside and outside an organization. Actors implement the knowledge creating spiral shown in Figures 2.1 and 2.2 and create new knowledge while producing ICT dynamic capability through the interaction of these three elements. Below, I expand on these elements one by one.

ICT Application Capability

The meaning of ICT application capability is not restricted to the technological functions of ICT itself or the skills of actors operating ICT. While ICT is certainly a tool for actors to implement daily organizational activities and business processes, a more important point is that actors also use it to create new meaning and context relating to daily activities while interacting with the strategic and organizational contexts of their own companies. Then it becomes important for actors to improve and transform the organizational practice of ICT activities to create new business processes and product ideas, and to implement new forms of ICT use and business improvements from improvised learning (Orlikowski, 1996) at the worksite. Transformation of this kind of organizational practice is enacted not just through ICT's technological functions, but through actors' ICT dynamic activity. In other words, the resource of ICT by itself does not determine an organization's social change. The ICT activity structure is determined by the planning and judgments of human agents (actors), and social change is implemented, and new structures formed, through actors' trial and error (Orlikowski, 2000).

Put another way, ICT activity as a result of specific thinking and behavior through actors' recursive practices revolutionizes routine and habit, and creates structures as new social systems (Orlikowski, 2000). The resonance of value (Kodama, 2001a) among actors rooted in a company's or organization's vision and mission provides meaning, power, and motivation aimed at actors' transformation. Then actors deliberately or emergently exploit ICT, and go on to repeat and create (production) routine change and new business innovations (creation of new structures).

From the viewpoint of agency perspective (see, e.g., Barley, 1986; Orlikowski, 2000; Orlikowski & Barley, 2001; Robey & Sahay, 1996), ICT's activity formation is socially constructed. ICT is created from new, diverse social meaning among actors, and becomes an "interpretatively flexible" entity supporting actors' business action (Orlikowski, 1992, 2000). An organization's most effective process for using ICT tools is greatly influenced by the environment surrounding actors and the strategic and organizational contexts. The effective introduction of ICT depends greatly on the context of the organization, and the installation structure and results differ from organization to organization (Barley, 1986; Fuller, 2000; Robey & Sahay, 1996; Van Mannen & Yates, 2001) as the ICT learning processes differ among organizations, even for actors using the same technologies.

It follows that different organizational settings lead to the enactment of different structural changes, even when the same ICT systems have been adopted (Barley, 1980). Put another way, the efficacy of ICT activities (its structures) created from the content of actors' thinking and behavior also differs among each actor (each organization). Looking at the VIN tools usage method in Chapter 1, Figure 1.17, the method differs according to a company's industry and conditions (including such factors as education, medical care, welfare, consulting, monitoring, and relaying) even for VIN tools possessing the same technology functions. Actors' usage habits and business structures also change according to intention and strategic objectives.

The introduction of ICT can also produce results different from the intentions of the original system designers as a result of an organization's strategic and organizational context (DeSanctice & Poole, 1994). Even though an organization may possess high-level ICT, positive, business-efficient results will not occur if actors use the technology inappropriately. Structuration theory (Giddens, 1979, 1984) claims that structures cannot be enacted as the "technology-in-practice" (Orlikowski, 2000) of rules and resources. The capability to use ICT also involves the specific thinking and behavior of actors who can enact new technologies-in-practice.

This kind of ICT application capability can be thought of as being constructed through organizational learning (Robey, Boudreau, & Rose, 2000) as a result of actors' trial and error with ITC. The mechanism of this construction process is described from the human agency perspective, Giddens's (1984) structuration theory (DeSanctice & Poole, 1994; Orlikowski, 1992; Orlikowski & Robey, 1991), and the actor network theory (Jones, 1999; Walsham, 1997), but the basic thinking is that its essence lies in the organizational learning process through the interaction of human agencies and institutions mediated by ICT.

As I mentioned at the start of the chapter, actors' minds, or specifically, actor's "productive interpretation" and "creative realization," with regards to ICT, can be considered important elements for practically enhancing ICT application capability. While the acquisition of ICT use capability can be considered separate to the strategic or organizational context surrounding actors, an improvised learning process of reinvention (Boudreau & Robey, 2005) arising, for example, from actors' ICT inertia can lead to productive interpretation or creative realization with regards to ICT.

The thought and behavior of actors attempting to embed ICT within a company's strategic and organizational context is an important element of their productive interpretation. Moreover, creating the inspiration of new knowledge from the interaction, in virtual space (ICT tool activity), of actors from different specialist fields is an important element of actors' creative realization. The significant elements existing in the background of this productive interpretation and creative realization (mentioned in Chapter 1) come to rely on management leaders' new perspectives with regard to ICT.

Management leaders have to consider new interpretations of matters from a variety of viewpoints. They must do more than weigh up the benefits and disadvantages of ICT's investment costs and technology features. They also have to consider a large number of themes, such as the impact of ICT on the company and society as a whole, the integration of the business process in-house, and the opposition from the organizational culture. Next, they have to accurately and objectively analyze the gap between current ICT strategies and future scenarios, and make an overall judgment on the benefits and disadvantages of ICT. They must be wary, however, on settling for a logically analyzed compromise plan. From the viewpoint of the agency perspective mentioned above, actors' dialectical action with regards to ICT creates a new perspective to be grasped constructively and productively. The new perspective of the management leaders then brings forth productive interpretation and creative realization from actors in the company, including the leaders themselves.

This kind of productive interpretation and creative realization is born of actors' new context architect and boundaries consolidation capabilities, as I will explain next. Then, in order to respond to dynamic changes in the environment, or to create change in themselves, actors promote the interaction of practice (through ICT activity resulting from the induced productive interpretations and creative realizations) and a company's constantly changing strategic and organizational contexts (see Figure 2.5).

Context Architect Capability

An important element of the second focal point of ICT dynamic capability is actors dynamically creating new contexts through interaction with ICT. Actors transform existing contexts and create new contexts through ICT activities (or through the medium of ICT). In other words, actors carry out business context architecture through the medium of ICT. In this book, I call this an actor's "context architect capability." I would like to take up this context architect capability as an example of part of the new product development process using empirical studies that I have reported on in the past (Kodama, 2007a).

Figure 2.4 shows the chronological process of a joint development between a Japanese communications carrier and an Israeli manufacturer. In step 1, discussions about the technical specifications of joint product development are carried out in Tokyo, and opinions are aligned on the meaning of development, specific technical specifications, and development schedules. In step 2, the Japanese and Israeli sides separate, activate VIN tools, and make progress with discussions on more detailed content specifications. In step 3, the actors transfer the location to Tokyo and have a technical meeting to determine the final technical specifications. In step 4, a prototype test of a software target product is implemented through

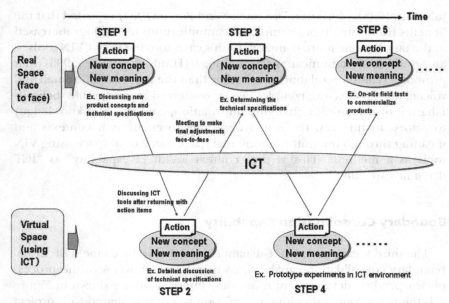

Figure 2.4. Interaction Between Real Space and Virtual Space: Case of New Product Development

international telecommunications circuits. The various themes developed up to this time are extracted, and two-way action implemented to solve problems. In step 5, field tests begin on the commercialized product with conditions established in Tokyo. More software-based problems emerge. After this, software debugging is carried out remotely from Israel over a broadband network, and the product is brought nearer to completion. Finally, development of the target product is complete.

With this example, complex situations emerged with various problems and issues at each step. Actors used VIN tools effectively to closely share information and context, however, and were able to create new contexts and link this to specific actions. Various novelties required to drive the project and solve knotty problems with the aim of new product development require strong purpose for what are, for the actors, the creation of new context and knowledge in a major situation (see the pragmatic boundaries in Figure 2.3).

In this case, the combination of face-to-face interaction in real space and interaction with virtual space using VIN tools raised the performance of the joint development project transcending different companies. Some hold the view that this is also consistent with past research results showing that a combination of media including face to face contact outperformed one without face-to-face contact (see, e.g., Ocker, Fjemestad, Hiltz, &

Johnson, 1998). Moreover, DeSanctice and Poole (1994) reported that the benefits from using more complex communications technology increased as the task became more complex. In this case, too, the use of VIN tools as an advanced communications technology (Hinds & Kiesler, 1995) that promotes real-time collaboration, rather than the non-real-time email and voicemail, to execute complex tasks conferred considerable benefits (sharing of dynamically changing information, contexts, and knowledge) to actors. In this way, the actors were able to create new contexts and meaning through the traffic among real space and virtual space using VIN tools as a medium. This is the "context architect capability" as "ICT dynamic capability."

Boundary Consolidation Capability

The third focal point of ICT dynamic capability is the element of actors' boundary consolidation capability. As mentioned above, with the process of new product development using the empirical studies shown in Figure 2.4, the new product development mission was deeply shared with project members. Actors faced the big decision of whether to develop with the guidance of their own company or develop jointly through a strategic alliance. The actors' decision to opt for joint development with the world's most advanced partners created new meaning and context and temporarily built an intercorporate network aimed at joint development crossing heterogeneous corporate boundaries. Then the new meanings and contexts created improvised action in the actors (Kodama, 2005a, 2005b). These improvised actions led to communication and collaboration with actors at companies holding heterogeneous values.

Although communication with actors belonging to different organizations (companies) and possessing knowledge from different specializations naturally gives rise to abrasion and conflict, the creation of new knowledge through actors' dialectical thought and action also becomes possible (Kodama, 2005a, 2005b). In this case, the combination of face-to-face interaction in real space and interaction using VIN tools in virtual space enhanced the performance of the joint development project transcending the knowledge boundaries of actors with heterogeneous corporate organizational boundaries and knowledge.

The above three elements of ICT application, context architect, and boundary consolidation capability form the core framework of ICT dynamic capability. These three elements have an interdependent relationship. Actors' context architect capability enhances their boundary consolidation capability. Then this newly acquired boundary consolidation capability creates new meaning and context, and further enhances context architect capability. This is a relationship of dialectical, recursive interplay.

Meanwhile, actors' inducement to productive interpretation and creative realization regarding ICT enhances actors' ICT application capability, creates new meaning and context among actors, and simultaneously enhances context architect and boundary consolidation capabilities through Figure 2.2's community knowledge creating cycle moving between real space and virtual space (see Figure 2.2).

Furthermore, actors' high-level context architect and boundary consolidation capabilities create meaning and context regarding ICT's new methods of use, and further enhance ICT application capability. Actors interactively exploit these three capabilities while spirally implementing the community knowledge creating cycle (explained in Figure 2.2) and creating new knowledge. By interacting with dynamic strategic and organizational contexts, actors implement the community knowledge creating cycle in spiral form, and simultaneously enhance these three capabilities in response to changes in the environment structure while creating ICT dynamic capability as an organization. The result is the creation of new business and construction of new business processes (see Figure 2.5).

High-level ICT dynamic capability enhances the operational efficiency and productivity (termed "knowledge efficiency" or "knowledge productivity" in this book) of corporate activity due to the construction of new business processes. This capability also enhances a company's creativity

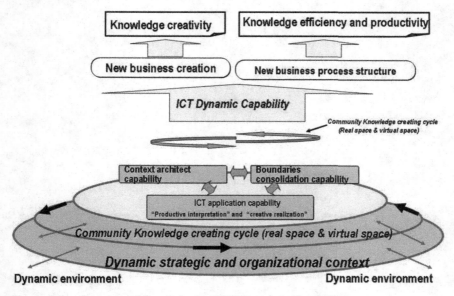

Figure 2.5. Conceptual Framework of ICT Dynamic Capability

("knowledge creativity" in this book) and creates a copy-resistant competitive edge inherent to a company, comprising new product and service development and the creation of new business models (see Figure 2.5).

NOTES

1. "Dynamic capabilities" are defined as organizational processes by which members manipulate resources to develop new value-creating strategies. Teece, Pisano, and Shuen stated that competitive advantage is achieved by a firm's "ability to integrate, build, and reconfigure internal and external competences to address rapidly changing environments" (1997, p. 516). In this book, "ICT dynamic capability" means the process by which companies form unique organizational capabilities conferring competitive superiority through the dynamic interaction of ICT-related resources and resources (knowledge and competences) inside and outside the company. The theories of Teece et al., however, do not indicate the detailed processes by which an organization accomplishes dynamic tasks of strategy formulation and implementation at a work-duty level.

2. The traditional information processing model (see, e.g., Anthony, 1965; Jonscher, 1994) uses the terms "information handling" and "information manipulating." In this book, however, I would like to use the terms "knowledge-handling model" and "knowledge-handling process" to indicate not just information but the higher-order knowledge concepts, including tacit and explicit knowledge.

CHAPTER 3

NETWORK STRATEGY AS PRACTICE

In this chapter, I look at how corporate actors acquire ICT dynamic capability through daily organizational practice. To investigate this, I would like to consider the processes of the actors' thoughts and actions from the viewpoint of the relationship between the corporate strategy-making process (Mintzberg, 1978; Mintzberg & Walters, 1985) and implementation. Through practice, actors come to recognize ICT tools as important business process tools aimed at the implementation of corporate network strategy.

First, I describe the positioning of VIN tools (a leading area of ICT) as strategy tools for corporate networks. Then I consider the question of how the ICT implementation process builds ICT dynamic capability as a result of different strategy-making processes within the organization. I will do this from the viewpoint of communications, context, and knowledge creation among actors.

VIN TOOLS AS NETWORK STRATEGIES

A great many companies are currently embedding all kinds of groupware and ERP tools, promoting communication and information sharing among organizations, and implementing best practice. The issue arises, however, even for companies that have already implemented changes in the corporate environment by embedding these tools, that email and groupware alone cannot sufficiently strengthen communication or speed up decision making inside and outside the organization. Some manage-

New Knowledge Creation Through ICT Dynamic Capability, pages 59–74
Copyright © 2008 by Information Age Publishing

59

ment and information leaders are voicing the opinion that rapid judgments and decisions are difficult to make when communicating or negotiating through unidirectional, non-real time media such as email.

The reason is that empathy and sympathy among individuals (employees) through face-to-face communication is a fundamental element at the heart of organizational communication. The opinion has grown that VIN tools are the ultimate practical ICT tools to completely resolve the closed-in feeling from email's and groupware's inability to sufficiently communicate individual thoughts and feelings, and to lead the way to corporate change. Put another way, these VIN tools enable content-rich discussion and swift decision making.

The growing need for VIN tools goes beyond the resolution of that closed-in feeling. For business in the future, various communities of practice and strategic communities (SCs) with external partners (including customers), such as strategic cooperation among companies, strategic outsourcing, and M&As, will continue to form. VIN tools are network strategy tools for the creation of new knowledge with these communities of practice and SCs.

Between 1999 and 2006, I surveyed 200 cutting-edge companies in Japan, the United States, and Europe (60 IT, 40 manufacturing, 40 distribution, 20 finance, 10 advertising, 20 education-related, and 10 medical care–related companies) that had already introduced VIN tools and dynamically exploited individual business processes, such as planning and development, design, blueprints, manufacturing, sales, distribution, and support. Figure 3.1 shows adjusted survey results from questionnaires and

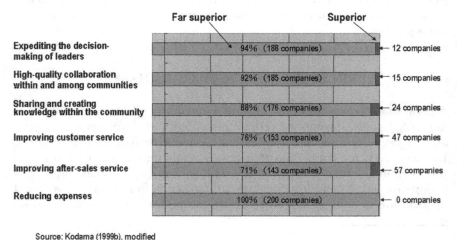

Source: Kodama (1999b), modified

Figure 3.1. Superiority of VIN Tools Over Groupware and Email

interviews with top-level managers equivalent to general managers, project directors, group leaders, and team leaders in the strategy, development and operation, service, and other divisions.

According to these surveys, VIN tools can be seen as superior to conventional email and groupware tools in promoting strategic activities within enterprise management. First, they are superior as management innovations to promote strategic business by speeding up management leaders' decision making; raising business efficiency from high-quality communication and collaboration within and among organizations comprising interaction among individuals; and sharing and creating knowledge and expertise in business communities.

Second, VIN tools are superior for enhancing customer service in the business community of companies and clients and for creating customer value with the marketing strategy aspect of comprehensive after-sales service. In this way, VIN tools enable judgments that make it possible to take on the roles of supporting corporate strategic activity and speeding up management innovation. In reality, we often hear the VIN tool user praising its efficacy by saying, "It's a tool that gets more interesting the more you use it," or "I can't work without this tool!"

I have had practical experience of and conducted many experiments in e-learning, such as employee education for companies using VIN tools. As a result, I have come to recognize the academic effect (improved skills) of satisfactory training (Kodama, 2001b; Kodama, Ohira, Kawakami, Kaneko, & Suzuki, 2004) even in the areas of home learning and gathering in satellite classrooms. VIN tools contribute greatly, even enhancing the knowledge assets of employee capabilities through remote training.

The basis of VIN tools is more than just a mechanism for efficiently distributing information held solely by individuals or companies. It also comprises active functions for information and knowledge to create new value in individual lifestyles and corporate activity, and to stimulate society and the economy. VIN tools for these purposes are systems expected to take on important roles as ICT platforms that create new knowledge and values on a network distributing not just text, but multimedia information including voice and images as well as diverse individual knowledge and expertise.

BOUNDARY COMMUNICATIONS

As I mentioned in Chapter 1, many examples of VIN tool activity are in such areas as conferences, monitoring of shops and other areas, and educational training. However, VIN tools display their true power when they are active as "boundary communications" among actors belonging to different organizations and possessing different specialist knowledge. In a dramatically changing competitive environment, companies in every indus-try must search for new business models while focusing on core compe-tences (Hamel & Plahalad, 1994) in order to build competitive excellence that other companies cannot easily copy. In particular, companies must for-mulate corporate strategy aimed at integrating diverse knowledge distrib-uted inside and outside the company (see, e.g., Kodama, 2007a, 2007b), especially while enhancing the importance of fusing different technologies and diversifying business models. The process of this knowledge integra-tion (this can also be the knowledge creation stage for the community knowledge creating cycle in Figure 3.2) is also an important theme for companies.

At the same time, moreover, the acceleration of Internet business, the specialization and division of technology domains, and the division of labor within an organization add further levels of complexity to boundaries of specialist knowledge among organizations inside and outside the com-pany, and among actors. However, as mentioned in the previous chapter, VIN tools become a highly useful practice tool to help corporate actors progress in removing boundaries, creating mutual understanding, and speeding up decision making among organizations and specialist knowl-edge domains.

But that is not all. Organizational and knowledge boundaries are places of innovation as well as of friction, confrontation, and contradiction

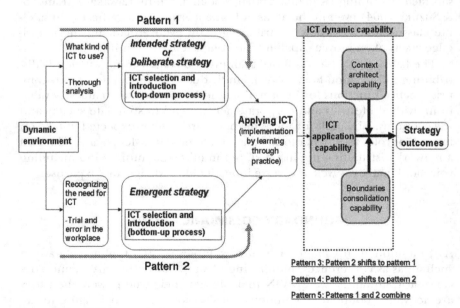

Figure 3.2.   ICT Dynamic Capability Through Network Strategy

among organizations (Leonard-Barton, 1996). Put another way, activating boundary communications can also create a driving force for generating creative technologies and new products. In fact, the introduction of VIN tools by a certain manufacturer adds confidentiality to boundary communications among organizations comprising marketing, development, production, and sales. In some cases, this has strengthened the process of product development.

So the question is whether organizational practice is necessary to implement VIN tools and generate "ICT application capability." After conducting interviews at a large number of companies and incorporating the results of my own action research, it became clear to me that the key enablers at the core of ICT capability of use were diverse communication among actors, the diverse context that it sprang from, and the management of strategy-making processes and implementation related to these elements.

BOUNDARY COMMUNICATIONS AND STRATEGY-MAKING PROCESSES

An important point of focus when implementing ICT is that changes in corporate business processes and organizational culture occur simultaneously (Brynjolfsson & Hitt, 1995, 1996; Davenport, 1993). Because of this, management leaders need to take both a top-down and bottom-up approach. Changes in the corporate landscape should use a top-down approach. Some companies, for example, have banned business trips for conference purposes on principle while simultaneously introducing VIN tools, which has had a great effect on speeding up decision making while restricting travel expenses. What is important in these cases is for the management ranks, and especially the CEO, to propagate opportunities for VIN tools use within the company.

The future will bring enlightened ICT use within companies, and an atmosphere where employees cannot function well without mastery of ICT, especially VIN tools. Banning business trips alone would probably have caused a reaction among employees. Because the use of ICT changes the corporate environment, however, the sense of having conferences away from the office as a "just cause" naturally diminishes, and the effect is to kill two birds with one stone by speeding up decision making and curtailing costs.

With the bottom-up approach, meanwhile, the actors in each workplace organization hope to build human networks that transcend organizational boundaries. Accordingly, common aims and values are shared, and dynamic communities of practice and SCs formed within the company.

The formation of Internet communities transcending physical space through the dynamic use of VIN tools enhances creativity generated from new products and technologies, and also bears on enhancing a company's competitiveness.

Network strategies for exploiting ICT in this kind of company, comprising the processes of deciding to introduce, introducing, and using ICT generally follow five patterns. Pattern 1 is a top-down approach, and pattern 2 a bottom-up approach. Pattern 3 comprises a shift from a bottom-up to a top-down approach, and pattern 4 a shift from a top-down to a bottom-up approach. Pattern 5 is a mixture of a top-down and bottom-up approach. Now I consider these management approaches in order.

The Top-Down Approach (Pattern 1)

The top-down approach is a strategy to implement ICT adoption analytically, rationally, and systematically in response to dynamic changes in the environment, with the aim of transforming a company's business processes or promoting e-business. Generally, the head office (whose mission is monitoring), the business head office's management and planning division, and the information system division demonstrate leadership, and strategy is formulated and implemented with the guidance of top management, including the CIO (chief information officer) and middle management among head office staff. Managers in charge of ICT especially need to go beyond analyzing the current business structure to enhance scenario planning analysis, and accurately formulate changes to the coming business structure and the new value chain projections.

From the viewpoint of the strategy-making process, the analytical, systematic, and rational introduction of ICT, centered on top management, is interpreted as implementation of deliberate or intended strategies (Mintzberg, 1978; Mintzberg & Walters, 1985). The top-down formulation and implementation of deliberate strategy creates shared context through univocal boundary communications (communication spanning various organizational and knowledge boundaries within a company), and builds a uniform strategy perspective within the company. Then the "equivocality" inherent in the various interpretations of actors within or among companies is reduced (Daft & Lengel, 1986), and strategy is systematically, rationally, and efficiently implemented through the deliberate approach of top management leadership.

"Equivocality" relates to various contexts emerging from diverse information and communication among actors. According to social communication theories, communication among actors includes the elements of univocality and multivocality (Wertsch, 1991, 2000). Univocal communication among actors communicates information as uniform meaning and forms shared recognition and viewpoints among actors. Set against this, multivocal communication generates diverse contexts and meaning, and heterogeneous, equivocal interpretation among actors generates friction and contradiction.

Equivocality exists alongside individual actors giving a range of contradictory interpretations (Daft & Macintosh, 1981; Weick, 1979). Multivocal communication has a high degree of equivocality, and arises from friction and conflict caused by differences of opinion among actors. Much equivocal information and interpretation within companies creates confusion among actors in an organization. Accordingly, companies have to reduce the equivocality in interpretations among organizations and actors relating to important decision-making items.

In order to reduce equivocality, top management and head office management leaders need to implement a constant flow of univocal data throughout the company, and create uniform contexts with the company actors. The creation of a strategy perspective unified through a top-down approach must reduce interpretative and contextual equivocality through boundary communications among actors within a company, and create a shift to unequivocal (interpreted in the same way as "univocal" in this book) boundary communications and contexts (Weick, 1979).

In creating a unified strategy perspective, top management and head office organizations need to process and analyze large quantities of information and extrapolate strategic data (a rational, systematic strategy). Then the management leaders need to simultaneously reduce the uncertainty and equivocality within the organization, and the actors in the workplace need to break down the strategic information to a level where they can act on the basis of specific indicators from the head office. With equivocal interpretations and contexts, it is difficult to share common strategy perspectives among actors. An example might be a negative reaction on the workplace side of the organization occurring from an equivocal interpretation of strategic content determined by decision making as a deliberate strategy of the head office. The introduction of ICT as a result of deliberate strategy can cause inertia among actors at the workplace organization. One factor for the failure of top-down implementation of ICT is the existence of these equivocal interpretations and contexts among the head office and the workplace.

The Bottom-Up Approach (Pattern 2)

This kind of bottom-up approach is a means of introducing ICT through trial and error, centered around workplace management, with the aims of improving daily business processes at sites close to the customer and improving and enriching customer services. Many cases exist of ICT tools being customized to improve (or transform) organizational practice among workplace staff. This trial and error process generates new organizational practices among actors. In many cases, actors in the workplace make minor or major changes to their initial plans or assumptions about the activity structure of the IT tools through trial and error and improvised action, and create structures such as new technology-in-practice (Orlikowski, 2000). It is not unusual, moreover, for people introducing ICT in the workplace to discover methods and effects of use that could not be predicted. From the viewpoint of the strategy-making process, this bottom-up approach is generally able to implement and interpret emergent strategies (Mintzberg, 1978; Mintzberg & Walters, 1985).

The process of trial and error to formulate and implement emergent strategies can also be termed a process of "abduction." Abduction is the process whereby managers in the workplace establish a hypothesis or model, and test and refine it through organizational practice. It follows that equivocality inevitably accompanies contexts and the formation of boundary communications among actors in the workplace.

The equivocality of these boundary communications and contexts creates pragmatic boundaries among actors, and friction gives rise to conflict. The trial and error process of abduction, however, creates the possibility of dialectically integrating the various contradictions generated among actors, and drawing out common strategic objectives. Then collaboration among actors aimed at implementing these strategy objectives is created (Kodama, 2007a, 2007c). As a result of this process, new meaning is formed from the common strategy objectives among actors.

A large number of companies with success stories of installing ICT in a single workplace organization find, however, that when they try to roll it out horizontally to other workplace departments, the actors in those departments do not necessarily interpret ICT in the same way (they might question, for example, whether ICT would be effective in their own workplace). It follows that implementing horizontal development of ICT throughout the company requires time and money.

Shifting from a Bottom-Up to a Top-Down Approach (Pattern 3)

The third pattern is a shift from a bottom-up to a top-down approach. Introducing ICT for the whole company during horizontal development, centered on workplace organization resulting from a bottom-up approach through emergent strategy, creates the issue of a higher burden of resources, time, and transaction costs. In this kind of case, however, top management from head office and management leaders lend significance to the success of an emergent strategy at the workplace, and the process of deliberately systematizing strategy becomes important. Then resource distribution resulting from the decision making of top management becomes an approach of justifying best practices throughout the company via emergent strategies in the workplace.

This kind of shift in management approach is often observed in large companies. Cases occur where ICT is introduced in certain branches and sales offices as examples for planning business improvements and better customer service. At large companies, for example, the head office organization leads by holding frequent meetings to present business improvements in the workplace. Managers from all branches and sales offices nationwide, or in each country around the globe, are introduced to the improvements simultaneously and given training or explanatory meetings. At this time, excellent examples of ICT activity are introduced to all participating members, and manager and head office staff in other workplaces acknowledge the reality for the first time.

Then managers at other branches and sales offices take action to study this example of excellence and introduce it immediately to their own workplaces, or else take another look at the possibility of adopting this ICT system in their own workplace. Meanwhile, emergent activities at certain workplaces create the possibilities of different interpretations (or of negative reactions) at other workplaces. In such situations, the formation of company-wide strategy perspectives relating to the introduction of ICT becomes important. This is the viewpoint of the strategy-making process shift from bottom-up, emergent strategy to top-down, intended, or deliberate strategy as a strategy concept of top management.

Because of this, top management divisions at head offices must redefine and adjust aggregates of diverse, equivocal strategy perspectives at diverse workplaces throughout the company, and build common strategy perspectives among members of the whole organization. The reason is that later reflection, adding meaning, and deliberate systematizing of the behavior of actors operating under the emergent strategy is important as a company's rational strategy. If the emergent strategy-making process is subsequently rationalized and justified, emergent strategies at the workplace, even if successful, cannot contribute to corporate strategy at all.

The common strategy perspective from the leadership of the top management at this head office requires common ownership of univocal boundary communications and contexts among actors within the company. The role of building joint strategy perspectives through these univocal boundary communications and contexts falls to the staff at the head office, represented by top management. The self-reflection (Shone, 1983) and "sense-making" (Daft & Weick, 1984; Weick, 1995) of head office staff discovers the newly consistent meaning of the company-wide ICT strategy implementation, and goes on to promote common ownership of company-wide univocal boundary communications and contexts. This company-wide ICT strategy becomes the trigger for the formulation of intended or deliberate strategies through a top-down approach.

When shifting from emergent to deliberate or intended strategy, the strategic concept that is the aggregate of diverse perspectives is redefined within the organization, and it becomes especially important to share ownership among the organization's actors. Close boundary communications spanning the head office and workplace sides accumulate new significance and understanding with regard to emergent strategy that workplace managers have come to promote. At the same time, the head office managers enable a movement toward corporate strategy as comprehensive and rational deliberate strategy embedding strategic concepts from the workplace, and an enhancement of the strategy's meaning. This further enhances the quality of a company's strategic concepts emerging from changes in this kind of strategy process.

Shifting from a Top-Down to a Bottom-Up Approach (Pattern 4)

This approach corresponds to cases where the top-down method of implementing ICT strategy formulated by top management and staff at head offices through pattern 1 could not obtain sufficient results. The primary reason is that the building of strategy perspectives through common ownership of univocal boundary communications and contexts as a result of head office leadership was unsuccessful due to deficiencies in communication among actors. The deficiency in communication is a factor behind the inability of managers and staff in the workplace organization to sufficiently draw out productive interpretation and creative realization, which are triggers for the ICT application capabilities described in Chapter 2. Conversely, boundary communications with moderate equivocality among actors, centered on workplace organizations, are required. Put another way, the strategy perspective of head office leadership that ignores the various interpretations of actors at the workplace organizations is a factor behind constraints on organizational learning of ICT applications in those organizations.

The second factor is business analysis among ICT's strategy formulation and implementation. Head office staff invests energy mainly in formulating analytical and rational strategy planning for ICT strategy, and planning implementation carries an assumption of workplace organization (power relationships among actors created from organizational hierarchies and apportioning of work duty roles). Accordingly, the level of boundary communication with staff at the workplace organization who mainly carry out strategy implementation is inevitably deficient, and the feedback route for ICT strategic policy from the workplace organization is automatically cut off.

Dialog as close boundary communication generates new context and meaning spanning different perspectives among actors. Dialog becomes an important element in knowledge creation (Nonaka & Takeuchi, 1995). To bring this about, the resource distribution of the deliberate strategy must be reviewed as a top-down approach, the authority of ICT strategy policy partially handed to the workplace organization, and CFT (cross-functional teams) and project teams built spanning head offices and workplace organizations. Then the top management must transform strategy through a process of trial and error resulting from emergent strategies centered on workplace organizations.

This example is described in detail in Chapter 7 as the case of Sony's introduction of ICT. Basically, however, management leaders need to temporarily curtail strategy implementation resulting from common ownership of univocal boundary communications and context from a top-down approach, and focus on planning to shift to a management style that prioritizes dialog through sharing equivocal boundary communications and diverse contexts centered on workplace organization. This will lead to diverse strategic perspectives among actors regarding ICT implementation, and emergent strategy will result in the creation of new forms of ICT use through the trial and error approach of organizational learning.

Combining a Top-Down and Bottom-Up Approach (Pattern 5)

The fifth implementation format combines pattern 1's top-down approach with pattern 2's bottom-up approach. From my past observations (including the results of my own action research) at the management workplace, I believe that this approach is the most effective method for implementing ICT strategy. Executing this pattern, however, requires commitment to ICT strategy formulation and implementation from middle management on the workplace side as well as from top management and head office staff.

From the viewpoint of the strategy-making process, top-down deliberate strategy and bottom-up emergent strategy means the coexistence of different strategies. Moreover, the coexistence of unequivocality (univocality) and equivocality is required as the structure for the boundary communications and contexts within the company.

The process of forming deliberate and emergent strategies involves mutually complementary dialectical relationships, not mutually exclusive relationships (Kodama, 2003a, 2003b). Realistically, planning division staff on the head office side not only analyzes resources and competence within the company and its environment when formulating deliberate or intended strategy, but thinks and acts with processes to implement rational decision making through trial and error. This trial and error process is one of formulation as emergent strategy. Moreover, although the trial and error processes of managers in the workplace organizations lead to the implementation of emergent strategies, the managers neither act freely nor ignore unified strategy perspectives, either as companies or organizations. Rather, they think and act deliberately and systematically while thrashing out comprehensive, whole-company contexts from the head office and the workplace organization's own strategies. In this way, very little purely deliberate or intended strategy and emergent strategy develops from managers' actual business activities. Rather, the elements of one strategy-making process become incorporated with those of another.

As with strategy-making processes, unequivocal (univocal) and equivocal boundary communications and contexts are mutually dependent. Even if boundary communications and contexts relating to clearly indicated deliberate or intended strategy are unequivocal (univocal), the fact that people attempt to analyze environments and resources from different viewpoints means these analyses include some equivocality prior to formulation. Similarly, even if equivocal boundary communications and contexts relate to the interaction of diverse dialog among actors in workplace organizations, when the actors attempt to reflect on action and create strategy concepts through that dialog, then unequivocal (univocal) boundary communications and contexts come into being as simple strategy perspectives among those actors.

THE ICT-DRIVEN COMMUNITY-BASED FIRM

As mentioned above, different strategy-making processes regarding the ICT implementation processes should be adopted depending on a company's strategic or organizational context. Deliberate or intended strategy is appropriate from the mature stage, when the initial stages of ICT implementation and the organizational culture of ICT use has filtered through to the employees. Emergent strategy is more appropriate, however, in cases where the process of forming the strategy targets productivity and efficiency from reform of existing business processes in the corporate growth stage, or searches (experience and incubation through trial and error) for means to create new e-business in the workplace organization.

In order for corporate ICT strategy to be effective in the long term, organizations at times need to shift from deliberate or intended strategy to emergent strategy (pattern 4). Conversely, as the opposite process, there are times when it is necessary to shift in the opposite direction (pattern 3). It also becomes important to implement a mixed pattern of both strategy-making processes (pattern 5).

Next, I consider the state of a corporate body that manages a flexible strategy-making process. Generally speaking, the mission of corporate activities must hold the long-term foci of achieving the expansion of existing business as exploitative activities while creating new business as exploratory activities (Kodama, 2007a, 2007b, 2007c). To bring this about, companies need to be supported by solid business visions pursuing value creation in different business domains. The essence of ICT strategy is to promote corporate innovation and new e-business while pursuing enhanced efficiency and productivity in existing business. Constant new knowledge creation activities aimed at realizing business domains based on this kind of ICT strategy mission must be implemented as corporate action missions.

The instigators who present the comprehensive concept of the ICT strategy and implement it strategically throughout the company are the top management of the head office and business offices as well as strategy and planning staff. They have to adopt rules-based, deliberate, systematic, and rational strategy plans based on a thorough analysis of the environment. They also have to make policies aimed at raising the efficiency of existing business and making incremental improvements, and plan ICT strategies aimed at new business (see pattern 1 in Figure 3.3).

In this sense head office organizations, which function as monitors, can be termed "deliberate organizations." The mission of these organizations must raise the unequivocality (univocality) of boundary communications and contexts for all organizations within the company, build common strategy perspectives among actors, and share them in the workplace organization and other areas. Core knowledge as ICT strategy generated from deliberate organizations is also explicit, formalized knowledge (called "planned knowledge" in this book) with a high level of completion formed from actors' unequivocal (univocal) boundary communications and contexts.

Meanwhile, the workplace organizations that are closest to the customer, such as branches and sales offices attached to business divisions and these organizational entities, must search for the most appropriate ICT strategy amid the practical processes of daily business activities. Naturally, the larger the corporation, the more the ICT strategy concepts are separated through

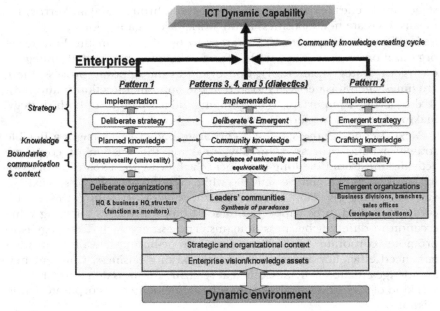

Figure 3.3.   ICT-Driven Community-Based Firm

business divisions corresponding to different business domains, and there are many cases of branches and sales offices where the concepts differ still further. There are also many cases where ICT strategy from deliberate or intended strategy formulated by the head office divisions cannot be said to be appropriate strategic content shared by all business domains. Accordingly, actors at workplace organizations must search for ICT strategy through emergent strategies by means of trial and error arising out of daily practical processes (pattern 2 in Figure 3.3). In this sense, workplace organizations can also be called "emergent organizations."

In a large number of emergent organizations engaging with business domains or customers, it is important for actors to share common strategic perspectives with each other. For actors in the workplace supporting actors' day-to-day business activities and customers, however, the strategy perspective is divided at the micro level, and the management leaders need to approve changes resulting from the day-to-day business context.

Accordingly, the elements of boundary communications and context equivocality inevitably loom larger at the workplace organization. Moreover, knowledge as ICT strategy generated from emergent organizations contains a large element of tacit knowledge formed from equivocal boundary communications and contexts (called "crafting knowledge" in this book). This crafting knowledge is elastically transformed by actors responding to strategic and systematic contexts at the workplace organization in a changing environment.

These two types of organization (deliberate and emergent) are generally classified with the paradoxical elements of pursuing efficiency and control on the one hand, and creativity and autonomy on the other, and tugs of war and conflict are always breaking out between the two types. These elements inhibit the shift to achieving a different strategy-making process (patterns 3 and 4) and the coexistence and integration of different strategy-making processes (pattern 5).

The shift to and coexistence of these strategy-making processes is promoted, however, by in-house leaders' communities (LCs). LCs comprise management leaders (CEO, executives, senior executive managers, branch heads, project leaders, managers, and other responsible parties) at each management level in the company, including top and middle management layers at deliberate and emergent organizations, and management teams, cross-functional teams, and task forces mixing the two layers. The LCs fuse and integrate organizations' individual knowledge (planned and crafting), and create new knowledge (this knowledge corresponds to community knowledge in Chapter 2).

LCs span deliberate and emergent organizations, involve actors in various departments, and initiate the community knowledge creating cycle (as explained in Chapter 2). Using LCs to implement the community knowledge creating cycle results in the creation of ICT dynamic capability for the entire company (see patterns 3, 4, and 5 in Figure 3.3). To realize ICT dynamic capability, it is important for LC management leaders to have thoughts and actions that can coestablish and synthesize creative and systematic strategy methods that seem contradictory at first sight.

Characteristic activities of the deliberate strategy-making process include analysis of the environment, whole-company decisions, and later, understanding of the actions that should be taken and the reality of the situation. Head office organizations need to avoid mixing strategies within the company and the process of trial and error. So doing, the boundary communications and context within the company will inevitably become unequivocal (univocal). The emergent strategy-making process, on the other hand, includes cases where actions that do not connect to the whole-company strategy are a behavioral feature. It is important, however, that management leaders think and act to permit uncertainty and failure from emergent strategies, evaluate the significance of the activities later on, and systematize the behavior of actors as a result of emergent strategy. It follows that different capabilities are required of organizations or their management leaders when implementing emergent and deliberate strategy-making processes.

The coexistence of a transcendent shift among the two strategy-making processes that comprise patterns 3 and 4 with the different strategy-making process of pattern 5 is essentially dialectic (see Figure 3.3). With dialectical thinking, we simultaneously accept the processes of forming deliberate and emergent strategies, which appear to be opposites. So doing, we attempt to integrate these processes by transforming and transcending them to a higher order. As with the boundary communications and contexts among actors, unequivocality (univocality) and equivocality—things that appear to be opposite—are accepted. It is then important to integrate these elements by transforming them to a higher order and transcending the framework.

The formation of these kinds of diverse strategy-making processes and diverse boundary communications and contexts leads to friction, conflict, and tension among actors. Among the opposing concepts, this friction and conflict can be thought of as tension between the dialectics of thesis and opposing thesis, and the integration of this tension leads to the achievement of a higher order, resulting in the building of a solid ICT dynamic capability. I would like to use the term "ICT-driven community-based firm" (see Figure 3.3) for this corporate entity that responds to dynamic environmental change (or self-generated dynamic environmental change) and generates ICT dynamic capability based on the dialectical thinking and actions of these management leaders.

CHAPTER 4

PROMOTING COMMUNITY MANAGEMENT THROUGH VIN

Amid recent changes in the business environment and progress in ICT, several companies are engaging in strategic business development by creating a variety of communities within and outside their corporate structures. Against the backdrop of these activities, in this chapter I explain the importance, given likely future developments, of community management for continuous activation of corporate organizations, and for new business creation. Then I present frameworks relating to community management using VIN activities at the company. The top-down approach from the innovative leadership of community leaders who comprehensively manage the business community distributed within and outside companies (including customers) will enable the dynamic application of community members' ICT. As a specific case study, this chapter describes applications for finance-related businesses and the automobile industry that use community management–supporting VIN, and points out that VIN will become an important broadband communication platform for business innovation.

New Knowledge Creation Through ICT Dynamic Capability, pages 75–96

THE FRAMEWORK OF COMMUNITY MANAGEMENT

Creating Strategic Communities

As mentioned in Chapter 2, one can think of companies as collections of collaborative communities composed of diversely populated teams or groups, with members from both within and outside the organization, that organically engage in strategic business activities and operations to achieve the organization's business goals. When considered with respect to companies' external environments, the communities within them can be broadly divided into three categories, as shown in Figure 4.1.

Type 1 communities are contained within the organization of a single corporation, and do work including routine, daily tasks, sharing information and knowledge for strategy development, and making decisions (moving from top stratum to mid-level, and then to lower levels; or from headquarters to branches and sales offices).

Type 2 communities are collective bodies formed with tie-up partners and outsource companies. These communities share diverse knowledge and information, create businesses, and perform work related to everything from outsourcing to strategic tie-ups.

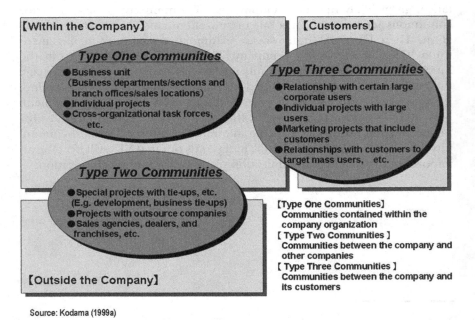

Source: Kodama (1999a)

Figure 4.1.   Community Types

Finally, type 3 communities are formed between companies and customers based on direct channels and new marketing strategies. New product and service development projects based on customer participation models that reflect the various needs and claims of specific customers are among the various forms these strategies assume. The customers range from large clients to general users.

To continually revitalize a company and nurture its efforts to create new businesses over the long term, it is important to create strategic communities of this sort. In this book, spiral business innovation through strategic community creation, in response to changes in the business environment and customer needs, will be referred to as "community management." Next, this chapter explains the essential factors that comprise community management.

Community Knowledge and Innovative Leadership by Community Leaders

There are two essential factors in strategically promoting community management. The first is the sharing, inspiring, creating, and accumulating of the diverse resources we will call information, knowledge, expertise, and ideas (community knowledge as described in Chapter 2). This applies to competencies possessed both by individuals and the entire organization, or community. Community knowledge can be defined as the power possessed by the overall community to incorporate resources (other community knowledge), within and outside the organization, in order to drive businesses as part of a creative new enterprise development effort having roots in business innovation by the middle and lower level employees who constitute members of the community. This power allows business results to be stored as diverse knowledge and expertise, and community knowledge to be formed continuously. An important point in this community knowledge formation process is the continual creation and creative development of community knowledge for strategic community creation, and the cultivation of exceptional employees.

The second essential factor is innovative leadership by the leaders within the communities. Community leaders, depending on the type of work they do, may be at the top (executives or other top managers) or middle of their company. The important point is that they demonstrate innovative leadership. Even if a company has the best possible community knowledge, that alone cannot guarantee success.

In other words, regardless of how many exceptional employees with how much ability a company employs, or however highly skilled they are, or how much marvelous, accumulated knowledge and expertise (community knowledge) the company possesses, there can be no hope of continuous corporate development and innovation without innovative leadership by exceptional people.

For example, if a community member suggested a product or service based on an exceptional idea, the proposal would not necessarily develop into a viable business without business mechanisms for effective promotion to the company's executives, customers, and the market, strategic outside political activity, and company-wide advertising and PR development. The missing ingredient is innovative leadership by a community leader.

Leadership, as used here, includes a variety of elements. The ability to produce, superior conceptualization and interpersonal skills, leadership, and political sense (both within and outside of the company) are especially important, however, for enabling a leader to make the thoughts and ideas of his constituent community members resonate to advance a common goal. A community leader must use his or her superior leadership to creatively develop and refine accumulated community knowledge, and must, through empathy and cooperation among leaders, smoothly promote all businesses the company should be targeting.

Strategic activities carried out personally by a community leader to implement his or her own new business proposal might include, for instance, making contact with other community leaders within and outside the company; vigorously engaging in external political activities when necessary; and taking the initiative to personally go to a customer's premises, assess his or her problems and needs, and conduct top-level sales. These sorts of actions on the part of the community leader have a great influence on his or her community's members, and may increase community knowledge. Given the dizzying changes in today's business environment and technological innovations in leading-edge fields, business strategy demands that community leaders, especially, think independently and take strategic action.

Promoting Community Management

This chapter has explained the importance of community knowledge in the area of community management, and of innovative leadership by community leaders that makes use of this knowledge to advance the goal of strategic community creation. In spiral business innovation through community management, it is essential that various strategic communities within and outside the company quickly grasp changes in the external environment, market structures, and customer needs by interacting and communicating on a personal level, and in so doing discover new information

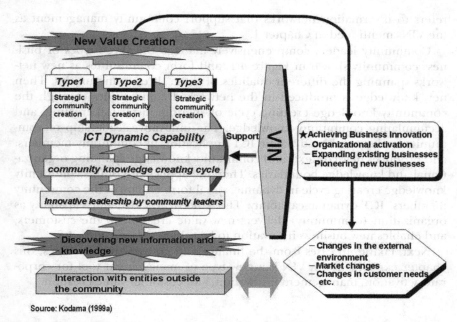

Source: Kodama (1999a)

Figure 4.2. Spiral Business Innovation through Community Management

and knowledge while at the same time storing and sharing this information and knowledge within the community as community knowledge.

The community leader must then make use of his or her innovative leadership to continually develop and redefine this accumulated, shared community knowledge, and thus create still more added value. In this way, continual organizational activation, the expansion of existing businesses, and new business creation become possible (see Figure 4.2).

By promoting community management, these activities bring about regular, personal interaction among the community leaders and constituent community members both within and outside the company, including customers. For instance, cases of new, personal interaction with a customer, or strategic tie-ups with various partners, triggering the creation of new products and services become possible. Contact with diverse communities and external knowledge also creates new value.

Given this sort of new value creation, it would also be possible, with community management that stresses personal interaction, to apply extra initiative and make effective use of ICT to achieve spiral business innovation through community management. And through this spiral business innovation, it would become possible to support the sharing, inspiring, creating, and accumulating of community knowledge, and to support innovative leadership on the part of the community leader. This chapter refers to information networks that support community management as the VINs mentioned in Chapter 1.

Community leaders comprehensively manage the three types of business community shown in Figure 4.1, and form communities as new networks spanning the different qualities of these three communities. Then new knowledge is produced in the networked communities through the community knowledge creating cycle of sharing, inspiring, creating, and accumulating community knowledge. The innovative leadership of community leaders enables dynamic ICT application in community members, and produces new contexts and community knowledge spanning organizational and knowledge boundaries. Then the application of this community knowledge creating cycle in dynamic, spiral form enhances the community members' ICT dynamic capability. This kind of ICT dynamic capability as organization (community) delivers new value creation to the customers, and enables new business innovation to be achieved (see Figure 4.2).

Next, taking up cases from the finance and automobile industries, this chapter explains how a VIN is a new type of management tool for corporate activation, management innovation, and customer value creation.

VIN IN THE WORLD OF FINANCE

The Need for New Management Support Tools

As globalization proceeds apace, the world finance industry is in the midst of a drive to acquire competitiveness, characterized by improvements in business efficiency, low-cost management, and the strengthening of retail sales. Retail sales techniques based on customer encirclement, especially, are attracting attention, necessitating new ICT strategies. Recently, retail business has been carried out aggressively via direct channels such as the Internet, mobile phone, and other network information technologies.

In the past, financial information systems focused primarily on infrastructure maintenance centered on administrative systems, including business systems (accounting), information systems (databases), and office work systems (sales outlets and operation centers). Now, new management innovation and customer creation management support tools have grown ever more important in creating new retail sales markets and achieving managerial efficiency and organizational activation in response to the globalization of the financial system.

Community leaders, especially those in top management, will not simply make rapid decisions in response to globalization and other changes in the external environment. They will also be required to respond to the shortening of product life cycles, to the creation of a greater variety of products in smaller quantities, to the tailoring of products to the needs of individual customers, and to various other changes in the market environment.

With conventional financial information systems, which focus on such processes as accounting, the gathering, processing, and analysis of standardized information needed for decision making is central. Compared with more strategic information systems that emphasize the environment for person-to-person communication between customers and bank employees (or among bank employees), these systems had a wider gap between functionality and applications. In other words, conventional systems were unequal to the task of accumulating and analyzing information and knowledge through the range of human interaction with customers, or for activating organizations in which human connections play a large role.

For example, for decisions involving important information and knowledge that can neither be standardized nor reduced to numbers, communication that includes persuasion, understanding, empathy, and cooperation among the various communities within and outside the company is essential. (See Figure 4.1, Types 1–3).

Increasingly important for expediting this new type of value creation through discovery and confirmation based on human interaction are not information networks based on the sharing of textual data via groupware and email, which has already become popular among various companies, but VIN tools composed of video-based, interactive information transmission networks (interactive video networks that use videoconferencing systems and videophone terminals, as mentioned in Chapter 1).

These VIN tools can, for example, support community management between customers and banks, among branches, between branches and sales outlets, between a company and its tie-up partners, between a company and its outsource companies, and in various communities formed through virtual corporations. VIN enables increased competitiveness and the creation of new business cultures.

VIN within and between Companies:[1] Application of Type 1 and Type 2 Communities

The long-established British bank, Standard Chartered Bank, with 600 business locations in 40 countries and 25,000 employees, has developed an international financial services business. It is a first-class bank whose international network ranks among the 100 most advanced in British companies. To continue improving its international competitiveness and lower its operating costs, the bank has installed a teleconferencing system in all its business locations, and is actively utilizing an internal VIN.

One striking practice of the bank is its active use of the VIN tool for meetings among management, financial experts, and its information technology group to achieve efficient communication and rapid decision making. Managers, financial experts, and other employees make use of the VIN to respond rapidly to changes in the political situation and economic trends in the United Kingdom and abroad. What's more, the VIN has made it possible for the ICT group to effectively monitor progress in the international network design and installation project. These are typical examples of Type 1 community applications.

As an example of a Type 2 community application, the use of VIN to consult with outside lawyers, via videophone, on matters ranging from general business dealings to formulating appropriate responses to problems is being promoted. The second example is Banco Bamerindus do Brazil, a large bank with 35,000 employees, 1,400 business locations in Brazil, and overseas locations in London, New York, the Cayman Islands, Luxembourg, Hong Kong, Venezuela, and other countries. The bank utilizes a domestic and international VIN to inform staff of new financial products, reassess existing products, and provide in-house training that allows its employees to upgrade their finance industry skills. Moreover, the system has made it possible for the top, middle, and lower levels of the bank, executives, and all sales outlets to engage in face-to-face, real-time, interactive, direct communication of top visions and ideas that are difficult to get across through email and similar tools. The system has also strengthened employees' sense of solidarity and created an environment where they feel more "at home" while at work. This is proof that a Type 1 community is contributing to motivating bank employees and activating organizations.

For the third example, both the Australian life insurance company G.I.O. Insurance, headquartered in Sydney, and Royal Insurance, a company headquartered in the UK that handles a variety of insurance types, are making use of interactive video communication with external sales agents to shorten the time required to decide important matters and close customer sales. These are both typical examples of a Type 2 community.

Shizuoka Bank, a top Japanese financial institution, responds to specialized inquiries on matters such as mutual funds by utilizing its network of videophones placed in all of its approximately 200 locations to connect those branches' customer service departments to specialists at the bank's headquarters.

On a larger scale, the global securities company Goldman Sachs has configured a broadband, high-speed, line-based videoconferencing system that reaches around the world (to the United States, the United Kingdom, Hong Kong, Japan, and other countries). Goldman Sachs executives use the network for rapid, worldwide, round-the-clock decision making on financial matters via high-quality video communication, and for the sharing of textual data.

Dealing and Trading:[2] Type 3 Community Application— Specific Customers

In an example of a Type 3 community that focuses on specific customers, the Allied Irish Bank Group, a large Irish financial services group, has linked its Dublin headquarters business office and its customer conference rooms in London, New York, and Singapore by means of a videoconferencing system in order to offer timely economic and financial information to its customers.

Over the videoconferencing system, using a combination of video and text sharing functions, customers are able to talk to Allied Irish Bank dealers and economic specialists without leaving their own offices, and thus enjoy the latest information. The system enables timely presentation of information and rapid response to inquiries from finance-related customers, creating a deeper level of trust among customers. The Bank of Boston has configured a similar video communication system, and is applying it in the area of financial trading.

Examples like these can be considered a new kind of marketing method that adds videoconferencing to conventional customer communication methods, such as telephone, facsimile, and email. They are typical examples of Type 3 community applications.

Multimedia Banking Kiosks:[3] Mass Users—A Type 3 Community Application

In an example of a mass user application, Type 3 communities are used for various types of virtual customer consultation, by way of videoconferencing systems and videophones.

A large American bank, Chase Manhattan Bank, has begun a trial service called Chase Interactive Banker that uses a videoconferencing system as a sales tool to affirm and expand its market presence. It currently offers the service at three of its business branches. Customers can talk directly with a bank representative over the videoconferencing system to apply for a new account or loan. A customer need only sit in front of a PC connected to the videoconferencing system to receive services with the sense of having made an actual trip to the bank. The system has the advantage of allowing banks to offer services by placing kiosk terminals connected to the videoconferencing system in small, ATM-equipped branches. This results in greater cost-effectiveness than if the terminals were installed in large branches.

US Bank also offers a video banking kiosk service it calls "ON-Sight." Using touch-sensitive buttons on monitors installed in kiosks, customers can establish face-to-face video connections with a customer service representative. Customers can now use the service for activities such as opening checking accounts, canceling overdrafts, and applying for credit cards, but US Bank plans to offer financial services such as consumer loans, home loans, and mutual funds in the near future.

This sort of kiosk service entails none of the administrative burden associated with establishing new branches or transferring employees, and it has the features of being easy to deploy in shopping malls and commercial office areas.

In Japan, multimedia consultation terminals have already been introduced, mainly by the consumer finance companies Acom Co., Ltd., and Promise Co., Ltd., and are in active use. Even more impressively, The Fuji Bank, Ltd., is introducing multimedia banks in Japan on an experimental basis, allowing customers to engage in activities ranging from detailed consultation to applying for loans, all by way of monitors connected to a video-conferencing system. Furthermore, as a commercial service, the bank is installing a new type of videophone-equipped ATM to provide financial product consultation and credit-card loan screening via videophone.

Moreover, the aforementioned Shizuoka Bank has installed multimedia ATMs, developed jointly with hardware and software vendors, in 35 convenience stores throughout Shizuoka Prefecture. These will provide a mechanism to allow customers to call an operator to their screen and consult on loans and other matters. The bank is also moving to strengthen its retail strategy in Shizuoka Prefecture to respond to globalization in the finance business.

The World's First Mobile Videophone "Visual Call Center"

One of Japan's leading consumer finance companies, Promise Co., Ltd., is installing the world's first "visual call centers" using mobile videophones (see Figure 4.3). These call centers enable applications and cooperation via mobile phone–based Internet access. Promise's visual call centers provide receptionists who impart a sense of security for new users making applications by videophone, and also enable established users to consult by videophone without needing to visit the bank. These functions are the VIN real-time functions described in Chapter 1. The weak point of the voice-only call centers operated up to now has been their inability to communicate content to the customers with detailed explanations. The visual call

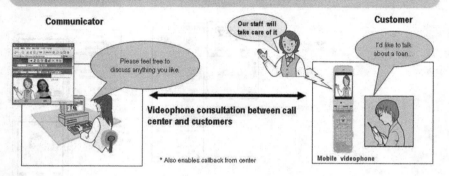

● Corporate image is enhanced by dealing with the customer sincerely through mobile videophone calls.

● Customers feel secure when they see the receptionist's face.

Source: Based on data provided by NTT DoCoMo

Figure 4.3.   Creating Customer Value Through Visual Call Centers

center enables simple, user-friendly explanations by enlisting the visual element to augment details that are difficult to communicate verbally.

The visual call center implements delivery of image data through mobile videophones (see Figure 4.4) as a non-real-time (storage) function of VIN. This involves attracting customers through original content, including visuals such as commercials and personality assessments. The call center also provides content necessary to undertake contracts, including methods and essential documents that are easier to grasp visually, and visuals of store information and ATM use that are essential to loans and repayments. In this way, the visual call center enables Promise to achieve greater efficiency and higher user satisfaction.

> In the last analysis, face-to-face communication is at the heart of service. The elements not just of high-tech but also "high-touch"—communicating with the customer—must be prioritized. The visual call center is essential for the future of our company as a channel connecting high-tech and "high-touch." (Manager at Promise Co., Ltd.)

Mobile phones are growing increasingly convenient as new channels for receiving services regardless of time or place. It remains true, however, that some customers prefer the sense of security from stores that enable them to meet someone face-to-face. VIN tools using mobile videophones are a new prototype for creating business that brings customer value from the formation of new communities with consumers.

◇ **Connectivity flow from mobile Internet**

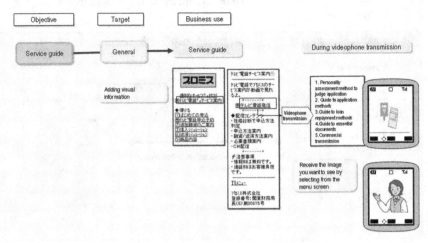

Source: Based on data provided by NTT DoCoMo

Figure 4.4.   Image Guidance Function of Visual Call Center

These are typical examples of Type 3 community applications focused on mass users. Much is expected from the application of Type 3 communities to the customer encirclement retail strategy.

A New, Integrated VIN for Finance Businesses

Core Concept and System

VINs built using video information transmission networks, and based on videoconferencing systems, videophones, and other terminals, differ from conventional, text-centered communications tools such as email and groupware in that VINs are primarily used for video and voice, but can also support the sharing of textual data when necessary. These capabilities enable the understanding, awareness, and empathy based on human interaction that are difficult to convey adequately using text alone, and also permit these qualities to lead to discovery, awareness, and creation of new values.

Despite all the talk about a headlong rush into the era of multimedia and ICT, human communication is not fundamentally one-directional and text-based (examples include orders sent by a boss to his subordinate via email, and notices and guidance sent from a higher-level to a lower-level organization). Indeed, the interactive video transmission of those elements conveyed through a person's manner of speaking, eyes, and facial expression are likely to grow increasingly important. Regardless of the extent to which technological innovation progresses, human communication will only become more important in the future.

VINs that support community management, as it relates to promoting businesses, comprise two core systems, as mentioned in Figure 1.8 in Chapter 1. These systems are divided along lines of real-time information transmission and non-real-time, storage-based information transmission, but share the three multimedia elements of image, voice, and text. Interactive systems are more effective than unidirectional transmission systems in carrying out smooth human interaction.

The first important core system for VINs that supports community management is a videoconferencing system (or videophone video terminals) that allow interactive, real-time video and voice communication as well as text, when required. These systems can truly be said to support the sharing of mutual understanding, decision making, ideas, and visions, as well as the new discoveries and value creation that result, based on interaction through human communication within and outside the community.

The second core system is Video on Demand (VOD), which is a storage-type video system that allows video and voice (as well as text, when required) to be stored, edited, processed, and played back. VOD systems are a tool that allows diverse community knowledge, within and outside the company, to be continually stored and enhanced. A storage-type video system is a database of knowledge stored as video rather than text and other conventional data formats. It is the most effective format for fostering community knowledge through organizational learning on the part of individuals and communities. A cutting-edge case in the consumer finance industry is that of Promise's VOD through mobile videophones, mentioned previously.

New applications for the finance business

Next in this chapter, I propose a new VIN application for the finance business that combines a real-time, interactive video system with a VOD system, and utilizes videophones and a videoconferencing system (see Figure 4.5). This section proposes a pattern of introducing videoconferencing systems and VOD into headquarters, branches, business offices, and other locations, and actively applying them to activities such as decision making by and among community leaders (top management, executives, project leaders, and others); decision making and the sharing of information and knowledge within and outside each community (among headquarters, branches, business offices, and overseas branches); and activities to cultivate personnel, such as in-house training. Such a pattern would enable organizational activation and rapid promotion of new ventures, based on

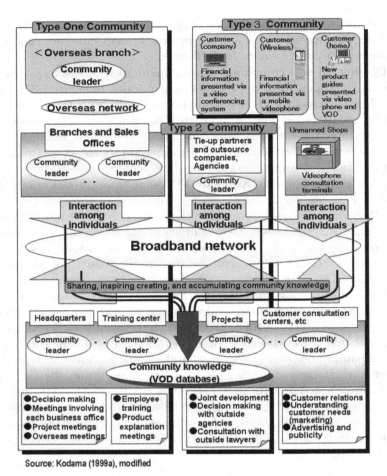

Source: Kodama (1999a), modified

Figure 4.5.   New Integrated VIN for the Finance Industry

interpersonal interaction within and outside communities and throughout the company.

Furthermore, it would be possible to use a VOD system to accumulate, edit, process, store, and play back the diverse information, knowledge, expertise, and ideas (community knowledge) that are created in each community, share and hand down this knowledge as corporate tradition within and outside the community, and in so doing start a new process of creating value for the company as a whole (sharing, inspiring, creating, and creatively accumulating community knowledge).

As an example of the customer encirclement strategy important in the retail business, I mention in this book the possibility of using home videophones (and mobile videophones when outside the home) to present new product information in video format. This could easily be realized using VOD delivered via videophones. The arrangement would also allow bank customer service representatives to respond face-to-face to various customer inquiries.

Using a PC-type videoconferencing system's telephone functionality or mobile videophone (as mentioned), a customer could enjoy the timely presentation of financial information through a direct video link to a bank's financial product specialist. Adding broadband connections, moreover, would allow customers to access financial information at any time. This sort of retail business activity enables new, unconventional types of marketing and the creation of new businesses that use communities to surround the customer.

Today, the household use of very low-cost, highly functional videophones as well as wireless networks through mobile videophones are growing in popularity at the same time as broadband networks are making large gains in the household and mobile phone markets. These developments mark a broadening of the world of communications from traditional voice- (or sound-) based telephones to video.

Viewed another way, this phenomenon suggests the possibility of creating various video-based businesses, including finance. With respect to educational and medical welfare businesses, it shows, particularly in the context of an aging society and shrinking number of children, that the application of VINs will continue to expand, as will the creation of new, virtual businesses through communities among different types of industries.

VIN IN THE WORLD OF AUTOMOBILES:
MANAGING WITH SPEED AND EXCELLENCE—
THE CASE OF PEREGRINE, INC.[4]

Promoting Community Management

Peregrine, Inc.,[5] headquartered in Southfield, Michigan, is an auto parts supplier and a first-rate company from the standpoint of quality control.[6] President and CEO Edward J. Gulda stresses his firm belief in the company's strategy of maintaining a competitive advantage through speed and excellence.

Behind his emphasis on this point is the reality that amid the diversity of customer needs and shrinking product life cycles that characterize his industry, it is becoming increasingly important for a company aiming to beat its competitors to invest in the capability to nimbly ascertain customer needs, and to quickly develop, produce, and market products.

The U.S. auto industry, which is characterized by major changes in the business environment, was subjected to an intense offensive by Japanese automakers in the latter half of the 1970s. As a result, companies in the United States vigorously promoted policies such as concurrent engineering (Clark, Bruce, & Fujimoto, 1987; Dyer, 1996; Ernst & Young, 1991; Helper & Sako, 1995; Womack & Jones, 1996). As automakers have efficiently applied these policies, suppliers have reduced the life cycle of the components they supply, in step with frequent model changes, from their previous lengths of 4–5 years to 2 years or less.

Consequently, suppliers must participate from the early stages of model development, so they assist in developing various components that fit their customers' needs through joint projects with their customers (supplier-maker design sessions). In this way they can greatly reduce the length of the product development and production process. So to aggressively promote concurrent engineering like that mentioned earlier, it is important to strengthen community management in relation to the three community types described in Figure 4.1.

As a Type 3 community application, the first form is community management between the automaker customer and the supplier (in this case, Peregrine, Inc.), which demands rapid decision making and concrete action in response to the diversified needs of the customer. For this to succeed, communication between the community leaders and community members of the customer and supplier, and the sharing, inspiring, creation, and accumulation of information, knowledge, expertise, and ideas represented by community knowledge, are always essential. Success may also require that the strategic level of top management exchange opinions and information and make quick decisions.

As a Type 1 community application, the second form is community management within the supplier's own company. An example is strong cooperation between the design department and the production department. Here, it refers to the smoothing of two-way communication and collaboration between the design and production group departments, in which it must be possible to share and create community knowledge among the groups, and to achieve rapid decision making by community leaders. Reexamining the production process and solving problems arising with changes in specifications from the automaker customer are among the important business processes within this community.

As a Type 2 community application, the third form is community management from the primary supplier as seen from the automaker (Peregrine, Inc.) and its own suppliers (second-tier suppliers). Specification changes from the automaker do not only affect the primary supplier, but have a large ripple effect on second-tier suppliers, which requires cooperation and agile community action to deal with effectively.

And so, in each of the three forms of community management, both within and among communities, implementing this management in a deft and proper manner is the most important issue for the purposes of reforming management and creating customers. Peregrine, Inc. strengthened community management by making a large investment in ICT to configure a VIN with the aim of further promoting concurrent engineering in the areas of auto design and production.

Reforming to Become a World-Class Company in Ten Years via VIN

VIN Configuration Using Intranets and Extranets

To smoothly promote communication and collaboration among its departments, the company used a high-speed, fiber-optic, wideband ATM network (ATM-LAN/ATM-WAN) to configure an intranet connecting its headquarters and four factories[7] (Ford, Lew, Spanier, & Stevenson, 1997). It went on to connect itself to its automaker customer and second-tier suppliers using ISDN, and an extranet was realized within each of these communities.

Meanwhile, as a concurrent engineering tool, Peregrine had been sharing blueprint files with its automaker customer over a large-scale CAD network to expedite design and development work. The company also achieved speedy management by adding face-to-face video communication and collaboration within and among the communities of the design staff, development staff, and production staff. Contributing to this is the high-speed, wideband, ATM network-based VIN the company has configured.

This VIN links roughly 900 customers (desktop PCs) in various departments (head office staff, design, and factory production departments) in five remote locations (including the headquarters) to an ATM network, and between 10 and 100 of these customers are equipped with a high-speed, high-quality videoconferencing system. What's more, these videoconferencing systems can also be connected to a public ISDN network to allow videoconferencing with the automaker and second-tier suppliers.

The VINs in each location are equipped with the standard features of a VOD system, allowing the diverse community knowledge created within and among communities to be stored in the form of video information.

Top-Level Sharing, Inspiring, Creation, and Accumulation of Community Knowledge and Rapid Decision Making (see Figure 4.6)

Mr. Gulda makes active use of the VIN by holding a strategy session involving business planning and financial analysis with his company's executive staff and factory heads every Monday morning from his desktop PC.

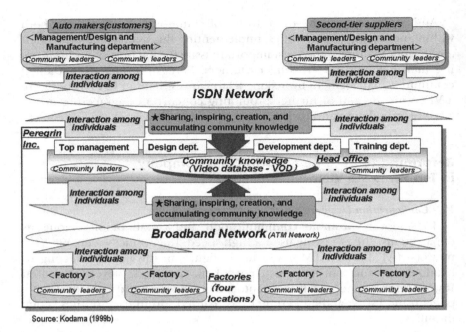

Source: Kodama (1999b)

Figure 4.6.   The VIN at Peregrine, Inc.

The VIN also plays an important role in face-to-face communication with customer automakers and second-tier suppliers. The top-level managers are always using the VIN to implement quick, strategic community management within and among communities.

Promoting Communication and Collaboration among Different Departments, Such as Design and Production (see Figure 4.6)

An important aspect of business operations in the intracompany communities is the promotion of community management using real-time communication and collaboration between factory and design teams that are approximately 60 miles apart. Designers and factory technicians use the community net to quickly solve current component problems in a virtual face-to-face setting. Furthermore, interactive, real-time collaboration between specialists from completely different fields, such as a door handle designer and sheet metal technician, is now possible.

Through this kind of community management using VIN, Peregrine can quickly offer superior products to the automaker customer, and the automaker can quickly place the end product on the market as a new model. This, in turn, has a direct bearing on improved customer service for the ultimate end-user consumer.

Storing Community Knowledge Using VOD (see Figure 4.6)

Peregrine uses a VOD storage system to continually share, inspire, create, and accumulate the community knowledge that arises within and among the communities of its organization. An important point relating to VOD is that it facilitates promotion of personnel cultivation, intracompany communities, and organizational learning for the entire company by allowing various educational materials for intracompany training as well as knowledge and expertise derived from the design, development, and production processes to be stored as video information. What's more, Peregrine is using VOD for storing and updating personnel information, and aggressively applying it to human resource management and other processes.

Top-Down Investment in ICT

To configure a large VIN such as that mentioned above, a network configuration professional with general knowledge is thought to require about 2 years. By using top decision making, however, Peregrine was able to complete the task in about 8 months. Behind this accomplishment was a firm belief at the top level that the VIN would lead the company to a superior competitive position.

A special project team composed of Peregrine's system integrator team and personnel from external vendors vigorously promoted the network configuration project. The company is further aiming for the future development of a state-of-the-art network in the United States.

What's more, Peregrine plans to expand its business overseas and develop its ATM backbone-based VIN to be the core infrastructure of its global strategy. Against the backdrop of these plans is a firm recognition by the managing staff, especially at the top, that communication and collaboration within and among communities is the true lifeline of Peregrine's business.

BUILDING ICT DYNAMIC CAPABILITY
THROUGH A TOP-DOWN APPROACH

Case studies of companies in the finance and automobile industries show that the encouragement of top management is the common aspect for knowledge gained with the decision-making process for ICT installation. Among these companies, overhead organizations (including top management) analytically, rationally, and systematically implemented ICT installation with the aim of promoting e-business (especially finance) or changes in their own business processes in response to dynamic changes in the environment.

Up to now, financial businesses have emphasized traditionally calculated, transaction-type ICT systems prioritizing data transmission and accuracy. Changes in the competitive environment, however, have led companies to prioritize a shift to a management style aiming to enhance diversity of product development and customer services, boost business process efficiency, and further strengthen the relationship between employees and customers. Now corporate activity to create communities giving rise to new innovations and new values vis-à-vis stakeholders within and outside the company (see Figure 4.1) have come to be emphasized. In other words, ICT systems as management innovation and customer value creation models have become important to the company (see Chapter 1, Figure 1.1). In the automobile industry, moreover, amidst a globally competitive environment, it has become increasingly important to strengthen the value chain comprising automakers, parts manufacturers, and customers aimed at establishing efficient production methods and strengthening product development. Installing cutting-edge ICT systems supporting parts development, total production systems, and supply chains to develop a wide range of product development have become priorities.

Accordingly, in the finance and automobile industries, it is important to analytically formulate ICT strategy and implement it throughout the company in response to dynamic changes in the environment. In these industries, the head offices (which mainly have a monitoring mission) and the management planning and information system divisions of the business head offices have displayed leadership, and the ICT strategies are formulated and implemented with guidance from top management. It is necessary, especially for managers supervising ICT, to go beyond analysis of current business structure to enhance analysis of scenario planning, and accurately formulate predictions of new value chains and changes that should come in the business structure. Organizational activity to analytically, systematically, and rationally develop and introduce new technologies (VINs using broadband and mobile networks) is also embedded by top management leadership. The delivery of financial products to consumers using mobile phones and the automobile industry's large-scale development and investment in ICT are leading examples of this.

Deliberate or intended strategies as ICT strategies centered on top management formulate community building and shared contexts through univocal boundary communications (communications spanning the various organizational and knowledge boundaries within a community) with organization members, and build unified strategy perspectives within the company. Then the various interpretations arising among employees within the company (equivocality) are reduced due to such deliberate strategies by top management leadership (Daft & Lengel, 1986), and strategy is implemented systematically, rationally, and efficiently. Establishing a unified strategy perspective leads to productive interpretation and creative realization of staff-related ICT strategy among employees within and among communities, and goes to enhance ICT capabilities of use for all employees. A still more unified strategy perspective simultaneously enhances boundary consolidation and context architect capabilities within and among communities, and goes to build ICT dynamic capability.

As the era of full-fledged networking approaches, to succeed with strategies that actively incorporate increasingly common new business styles, such as strategic alliances, outsourcing, and e-business, companies will require from community leaders superior leadership that results in the continuous creation of community knowledge.

To achieve this, it is important for companies to work toward management innovation and customer value creation models (as seen from company case studies) by dynamically reforming the individual and organizational cultures of community leaders and members, especially top management, through full application of the digital, multimedia networking environment of a VIN.

The goal of applying VIN as a tool to support future network strategies is not just the staid idea of trimming travel budgets. The true benefit of these networks lies in areas such as gains in time and value through rapid decision making by first-rate community leaders, who are no longer bound by time or distance. It includes the promotion of efficient business operations by employees that occurs when communities collaborate internally and externally, and the constant creation of community knowledge. These advantages relate directly to the realization of managerial speed and excellence, a customer value creation business model, and progress in community management.

NOTES

1. Interviews were conducted with supervisors at Standard Chartered Bank, Banco Bamerindus do Brazil, G.I.O. Reinsurance, Royal Insurance, and Shizuoka Bank. The case study of Goldman Sachs on NGI (next-generation Internet) applications was presented in the NGI Video Summit 1998 in San Francisco.

2. Interviews were conducted with supervisors at the Allied Irish Bank Group and Bank of Boston.

3. Interviews were conducted with supervisors at Chase Manhattan Bank, US Bank, Acom, Promise, The Fuji Bank, and The Shizuoka Bank.

4. Interviews were conducted with supervisors at Peregrine's headquarters and factories.

5. Peregrine, Inc. is a supplier of internal and external components to the Big Three auto manufacturers of General Motors (GM), Chrysler, and Ford, handling design, production, and sales. Its wide-ranging product line includes seat systems, door modules, door trimming and related hardware, airbag systems, instrument panels, window regulators, and steering wheels.

6. In 1997, Peregrine received the international quality standard QS-9000 certification in its core competency fields of design and manufacturing. The QS-9000 quality standard system was jointly established by the Big Three based on the international standard ISO-9001, with the addition of conditions unique to the automotive industry. And recently, Chrysler awarded Peregrine's Warren (Michigan) plant, a zero-defects manufacturing plant, its Gold Pentastar Award, given to recognize its best supplier.

7. ATM (asynchronous transfer mode) is a system for transmitting images, voice, and data using a high-speed LAN (local area network) and a WAN (wide area network) that connects users across geographical boundaries such as city and state lines (see Ford et al., 1997).

CHAPTER 5

THE PROMOTION OF STRATEGIC COMMUNITY MANAGEMENT UTILIZING VIDEO-BASED INFORMATION NETWORKS

This chapter describes the importance of strategic community creation as a new management style. It verifies that video-based information networks (VIN) utilizing information and multimedia technologies enhance the quality of community knowledge possessed by strategic communities, and it also verifies, through case studies, that these networks are valid as organizational learning support systems within the strategic communities. Innovations in the area of veterinary medicine utilizing VIN tools in Japan over the past 5 years are taken as examples. This chapter describes how community knowledge within strategic communities comprising industry, government, and academia is enhanced, how the new virtual methods of telemedicine and distance learning are incorporated into the business process, and how concepts of regionally linked cooperative bodies are realized.

New Knowledge Creation Through ICT Dynamic Capability, pages 97–124

INTRODUCTION

Rapid progress in information and multimedia technologies is leading the way for gradual renovation in diverse areas including society, economy, and industry. Ever widening acceptance of the Internet, intranets, and extranets is spawning the flattening of corporations on new communications platforms as well as the creation of new-business models for intercorporation transactions as championed by e-commerce. The Internet is poised to revolutionize how new jobs are handled by people in corporate settings and even the lifestyles of individuals in their daily lives. It is also stimulating the proliferation of SOHOs (small offices and home offices). New business styles based on such concepts as virtual teams and virtual communities are representative of such trends (see, e.g., Lipnack & Stamps, 1997).

Amid such contemporary changes, the advent of the new 21st-century networking generations will usher in major changes in the value systems of individuals, especially as they relate to living and working. At the same time, it is thought that in the years to come, increasing importance will be given to the manner of existence of, and the new ways of thinking about, the "communities" that are represented by corporate entities and non-profit organizations, and that comprise huge numbers of individuals. Several writers have already discussed concepts of building communities in the workplace and reforming business processes (Bechard, Goldsmith, & Hesselbein, 1996; Hesselbein, Goldsmith, Bechard, & Schbert, 1998; Wenger, 1998, 2000).

In corporate settings, in particular, the knowledge management method, refined through rapid IT sophistication, is being adopted to address internal particulars, giving rise to structural renewal in affected areas.

But the important point here is that, no matter how ICT is embedded for business-related innovation, a corporation's strategic behavior most importantly depends on innovation of the value systems of the individuals concerned and of the knowledge and core competencies they have accumulated. Leadership continuously generating business innovations that strategically tap the knowledge and core competencies of extracorporate human resources, including customers, will become important.

Toward such an end, it becomes most important for corporate leaders to aggressively create strategic communities, tapping their own organizations as well as outside contacts, including customers, for leading staff positions, with the aim of innovating their own in-house core competencies while simultaneously creating new values and offering them to their customers (Kodama, 2007b). Regarding strategic community creation up to the present, a report using examples of ICT innovation in the corporate body has recently appeared (Kodama, 1999b). An important issue is how to create a strategic community and achieve innovation. A key element in the process of innovation is to actively incorporate information and multimedia technologies into various business processes. It is important to promote reforms not only in the corporation but also in various communities (Kodama, 1999a).

This chapter describes the concepts of strategic community creation and verifies, through case studies, that VIN tools using information and multimedia technologies are an effective system to support the innovative activities of strategic communities.

STRATEGIC COMMUNITY CREATION AND VIN

The Importance of Strategic Community Creation

Companies must continually evolve by engaging in various forms of innovation. Particularly since the advent of the knowledge-based society (Toffler, 1990), businesses have been faced with a major transition, from focusing solely on developing products and services to engaging also in strategic innovation in order to improve their business processes.

Innovation is a process that can occur in the course of performing various business activities, such as product and service development, marketing, manufacturing, sales, distribution, and after-sales services. It occurs not only in the course of improving and expanding existing ventures, but also while creating new businesses and ventures (Kanter, Kao, & Wiersema, 1997).

For various large, cutting-edge businesses with core competencies in software, semiconductor, and networking technologies forming the foundation of multimedia and computer networks, as well as for numerous venture companies, the past several years have brought increasingly fierce competition to leverage such new business styles as strategic partnerships, mergers and acquisitions, outsourcing, and virtual corporations, for the purposes of developing strategic enterprises, expanding the market shares of their products and services, and creating new businesses.

The important issue in promoting this kind of business innovation is not how to carry out strategic business operations using company resources (including knowledge, competence, and personnel), but rather how to develop innovative businesses by creating communities of various organizational forms through collaboration (including virtual collaboration via ICT networks) with resources outside the organization.

To continually revitalize a company and nurture its efforts to create new businesses over the long term, it will be important to respond to changes in the business environment by developing through strategic community creation (including virtual community creation via ICT).

A strategic community as described in Chapter 2 refers to a variety of bodies or organizations from both within and outside the organization that can create revolutionary innovations through new products, services, or business structures.

Four points about strategic thinking and behavior are of importance to the community leader (in this book, I refer to leaders in strategic communities as "community leaders"). The first is that the community leader, who is a member of an organization, must be able to understand the exterior environment in which the organization is placed, the pace of technology, market structure, and customer needs quickly and interactively.

The second is that the vision supported should integrate the leader's thoughts and beliefs, and that the leader should be able to create person-to-person networks in terms of stated concepts, both within and outside the organization. The extent to which contact and constructive dialog occurs with other core leaders (within and outside the first organization) will depend on the leader's networking ability.

The third relates to the importance of creating an arena of sympathy and resonance for value outlooks respecting visions and concepts through constructive dialogues with core leaders both within and outside the leader's organization (Kodama, 1999b).[1] This, in turn, makes it possible to create a platform for harmonizing the value outlooks of core leaders, creating room for the birth of a strategic partnership with core leaders.

The fourth is to create an organizational strategic community. For these things to happen, it becomes important to create a value-harmonized platform on the level of an organized body within a strategic community, comprising multiple, heterogeneous entities, with the participation of many community members.

With this firm resonance platform as a base, the quality of community knowledge (Kodama, 2000),[2] which is the knowledge inside the community, is enhanced by the innovative leadership of the community leader. The community competencies (Kodama, 1999b),[3] which are the core competencies inside the community, are also enhanced. The series of innovation processes of the community knowledge and competencies creates new value, and the business goals in the strategic communities are achieved (see Figure 5.1).[4] This sort of strategic community creation is referred to in this book as "strategic community management."

The Organizational Learning Support of Strategic Communities Using VIN

In order for individuals in the strategic community to further enhance community competencies and knowledge as a group, the individual needs

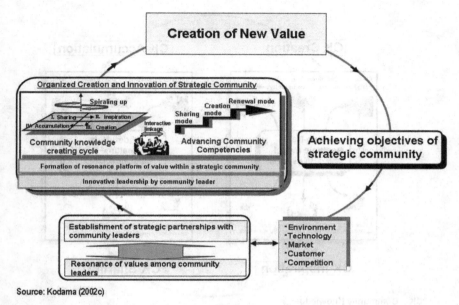

Source: Kodama (2002c)

Figure 5.1.   Strategic Community Management

to learn both as an individual and as a group community. Dialog and collaboration among individuals are necessary elements of this learning in strategic communities.[5] Individual community members thoroughly learn, understand, and seek out the mission and goals supporting the external environment in which the strategic community is placed. It is also important for the strategic community as a whole to generate power in order to achieve, through dialog and collaboration among community members, the mission and goals that the community has taken upon itself.

Building an ICT-based learning support system is central to assisting this kind of learning in the strategic community. As regards individual learning, for example, community members at all times need to gather and accumulate information and knowledge while seeking out, sharing, and understanding the information and knowledge of other community members. In other words, the learning support function of non-real-timeliness is a necessary prerequisite of the system.

Also required is a function whereby community members participate in real time or interactively; engage in dialog and collaboration with each other based on information and knowledge that each individual has shared and understood, inspiring shared knowledge and creating new knowledge as a strategic community; and organizationally acquire these items as expertise within the strategic community.

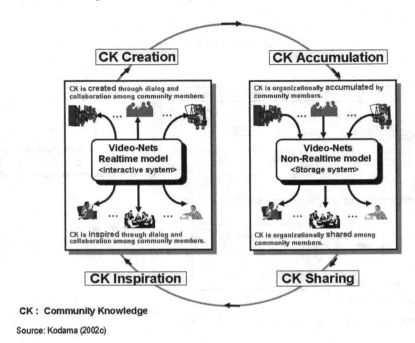

Source: Kodama (2002c)

Figure 5.2.　Innovation of Spiraling Community Knowledge-Creating Cycle via Video-Nets

In this way, a learning function of both real-time and non-real-time models (described in Figure 1.8 in Chapter 1) that also comprises an organizational learning support system is required to spiral up the innovation cycle of sharing, inspiring, creating, and accumulating community knowledge within the strategic community (as described in Chapter 2), and to enhance community knowledge as a whole while enhancing community competencies (see Figure 5.2).

CASE STUDY: INNOVATION IN THE FIELD OF VETERINARY MEDICINE USING VIN

Current Status and Issues in the Field of Veterinary Medicine in Japan

In Japan at present, there are 27,000 veterinarians, about 30% of whom are working in practice and research related to public health inspection, including the inspection of food items. Recently, new fatal infectious diseases such as the Ebola hemorrhagic fever have appeared on a global scale, and zoonotic infections (communicable to people and animals) contracted from animals and pathogenic microbes such as O-157 are crossing national boundaries, so that the role of the veterinarian is expanding.[6] Several years ago, health veterinary medicine departments in Japan were reorganized in research institutes, and it has been pointed out that teaching staff numbers at universities were about half those in Europe and the United States, with the result that research and education levels were falling behind. The European Union completed the standardization of common teaching criteria and qualifications for veterinarians in 1999, and the United States and Canada are also close to internationalization by standardizing their national veterinary examinations and qualifications. The internationalization of veterinarians' qualifications is also making progress in Japan, where it has been recognized that bringing veterinary medicine teaching levels close to international levels is a matter of urgency.

Unless the levels of education and research are brought into line with European and American standards, there is a serious fear that Japan's veterinary medicine will be excluded from the international fold. Under these conditions, international cooperation and exchange of information between veterinarians in the future will be essential. This is an issue common to industry, academia, and government that must be solved jointly.

In January 1996, Dr. Hirose,[7] Professor Emeritus at Obihiro University of Agriculture and Veterinary Medicine, located in Obihiro, Hokkaido (hereafter called the Institute), launched the Animal Medical Information and Science Development Research Institute as a joint venture of industry and academia. The research staff who comprise the Institute come from different working backgrounds and include university instructors, practicing veterinarians, group veterinarians, businessmen, dairy proprietors, and general citizens.

Giving greater significance to veterinary medicine in the community not only requires gathering the opinions of us veterinarians who specialize in veterinary medicine, but also preventing the narrow fixing of concepts by creating opportunities to open up experience and ideas with the participation of many people in other fields (different industries). In particular, when it comes to the prevention of disease to minimize economic losses with respect to production livestock, we can never expect true results without a practical plan that involves the participation of the producers (agriculturalists). In other words, it is not enough to have a veterinarian with abundant knowledge. Several veterinarians who can pool their knowledge will be required in order to dig up true value. Now is the time when our eyes must turn toward what the producers are earnestly hoping for: the activities of veterinarians with such flexible ideas as this. It is very important to exercise free thinking in order to break down inflexible veterinarians. If the conventional framework is not demolished when it needs to be, there will be no new ideas or personnel.

For interdisciplinary research, it is certainly necessary to set out a framework in the area of modern clinical veterinary medicine and to create an interdisciplinary organization from previously nonexistent collective bodies (collective bodies of different industries). There should be collaboration in creating together within the collective body, with each veterinarian developing their own original ideas.[8]

The aim of the Institute was to construct a virtual organization with 54 research members scattered throughout Japan, "a concept that gives full rein to independence in towns and villages, that makes manifest local expectations and characteristics, and is not centralized in Tokyo, the capital of Japan." An additional objective, from the international viewpoint, was that it would serve as a base for the transmission of information.

Meanwhile, little progress has been made in the preparation of a new media environment in the area of clinical veterinary medicine. Utilizing new media for education and research in veterinary medicine and the activation of local communities was one of the key points mentioned above.

Specifically, as explained in the next paragraph, the use of VIN tools enables the sharing and exchange of information on veterinary medicine through images formed in real- and non-real-time, which then greatly expands communication among veterinarians, and expands and establishes new virtual methods such as telemedicine and distance learning.

Creation of Innovation by Merging Different Business Areas

For some time past, X-rays, CT, MRI, and so on have been used in veterinary medicine on both large animals (such as cattle and horses) and small animals (such as dogs and cats), and are still being used for the imaging diagnosis of various types of disease, with great success in early treatment and prevention. In particular, the latest trends, involving efforts to reduce the cost of animal production as agricultural competition intensifies internationally and enhancing the status of companion animals in a household as people age and have fewer children, are creating a major flow from treatment to prevention in veterinary medicine as in the human world, and the role of imaging diagnosis is growing.

Among these moves, Dr. Hirose and other researchers have introduced Japan's first diagnostic X-ray vehicle dedicated to large animals to all universities in Japan, and are conducting numerous pioneering trials (Hirose, 1985). Nevertheless, the current picture is that the history of imaging diagnosis in veterinary medicine is still recent, and there are few specialists. This has made it necessary for distant veterinarians to collaborate in education, diagnosis, and treatment instruction in the process of developing and expanding imaging diagnosis.

Meanwhile, the barrier between human diseases and those of pets and other animals has been lowered, and the time has arrived when quick action through imaging diagnosis is required. On that point, the demand for remote, bidirectional treatment through imagery is growing, not only among veterinarians but also in areas of clinical veterinary medicine between veterinarians and animal breeders of all sizes.

Regarding other issues, uniform distribution routes to the production areas of beef cattle and food processing companies were built in response to social demands for more stable food supplies and higher productivity, disease prevention, and guaranteeing food safety. The need to contribute to food safety also led to the building of systems using video images to exchange information in the areas of controlling meat quality and gender conditions between distant locations, and to satisfy the need for a unified production and supply system. The need for producers, agricultural cooperatives, food processing companies, fertilizer manufacturers, research institutes, veterinarians, universities, and others to band together in order to control meat quality, perform inspections and analyses on fertilizer supplies, and discuss policies aimed at improving meat quality in response to demand has become an urgent issue.

On this basis, at the start of 1996, the research group centered on Dr. Hirose began research and development of remote veterinary diagnosis. In a partnership with company A in the private sector, the research group developed its own VIN tool in the field of veterinary clinical medicine. This system was built through strategic community creation in partnership with six project team members from Nippon Telegraph and Telephone Corporation (hereafter NTT) as a telecommunications company and 54 researchers from the research institute who participated in the development.

Bottom-Up Strategy for ICT Solution Proposals to the Customer: Implementing Emergent Strategies

At that time, NTT development project members were struggling against the odds with the development of VIN tools applications and customer proposals. At first VIN tools were introduced mostly as conferencing systems at general companies, but methods of use in specialist domains such as medical treatment and education had yet to be developed. The bureaucratically structured hierarchy of organizations at medical institutions, including general hospitals, ranges from university organizations and affiliated university hospitals at the apex to associated hospitals and then to medical clinics.

Up to that time, NTT development projects had repeated the top-down approach (proposals to the upper-layer management class of customers), starting with major hospitals and universities possessing medical institutions, when proposing VIN tools to customers. Development project members produced analytically disciplined proposal documents and carried out demonstrations to numerous customers. These seldom led to the expected result of orders, however.

Then project members changed their VIN tools proposal strategy. Rather than approaching the top management layer of large hospitals and universities, they began to seek out and promote communications with middle management layers and with doctors and medical staff at small and medium-sized hospitals that were greatly involved with VIN tools. Project members collaborated closely with users (who I will term "innovative customers" in this book), aiming to experiment through trials involving VIN tools in a major way, and acted jointly to seek out the benefits and disadvantages of the tools by trial and error. The veterinary medicine field possessed a similar hierarchical structure to that of human medical treatment, with universities and scientific societies at the apex. The project members' point of focus became how to discover innovative customers without the approach of making proposals to the top level of big universities. In this case, project members deepened communication and collaboration with the director of the small-scale research institute, Dr. Hirose, and implemented new, emerging methods of use for VIN tools in the domain of veterinary medicine.

When proposing ICT solutions to customer organizations, many businesspeople take the practical plan of approaching top management as the logical path. In other words, the client company's top management strongly recognizes the need to introduce ICT, and this is a process where client companies go on to introduce ICT at workplace organizations top-down to the entire company. This process can be interpreted as implementing deliberate strategy through the top-down approach of pattern 1, discussed in Chapter 3.

Set against this, the approach to middle management and lower levels at large organizations, or to small-scale organizations subordinate to large organizations, does not extend to taking the decision to implement full-scale ICT. Successful cases of local ICT implementation through the decisions of such lower-level management staff, however, can trigger dissemination to other organizations. This process can be interpreted as the implementation of emergent strategy as pattern two's bottom-up approach. The pattern of selecting successful models as best practice from a process of trial and error resulting from the implementation of this emergent strategy, and going on to develop the results for other companies, is also an important focal point. In this case, the team of Dr. Hirose (an innovative customer at a small-scale research institute) and an NTT project team formed a strategic community, aiming to develop a new VIN system using VIN tools in the field of veterinary medicine by trial and error resulting from emergent strategy, and to use this system to create a new business model.

This new VIN system was constructed on the basis of opinion summaries from the institute's 54 participating research members. The main objective was for all of them to be able to exchange clinical image information and knowledge easily on the same level. Specifically, it was a system structured to support the mutual exchange of information and knowledge with a uniform method, in which all research members would have the same equipment.

Developing and Constructing the New VIN

The first issue for the strategic community was the conception and design of the VIN. The information and knowledge possessed by community members was shared, and the constructive dialog between the many community members toward the design of the ideal VIN triggered mutual knowledge within the community. The factor that bonded each member to the community was the common value with respect to their vision and concept of "innovation in the field of veterinary medicine." In order to create and provide new value for many veterinarians, community knowledge was triggered and created by both the physical and intangible aspects of the VIN, including its conception, design, construction, and method of use, while bearing in mind what is meant by the application known as "interactive video communication."

A major problem occurred at the initial stage, however, arising from the use of personal computers by veterinarians. A PC is undoubtedly a convenient device, but this does not mean that it is always an excellent data system. A device with advanced processing capacity and a device that is easy for a person to use are separate things. The reluctance of the "informationally handicapped" to come forward made it difficult to match the PC to the person. Instead, it was necessary to construct a system equipped with the ideas and characteristics of the community and having functions for the exchange of information that anyone (including veterinarians of various ages) could use in a similar way to the household telephone. The community therefore had no alternative but to abandon the idea of applying the video-conferencing system, which uses a PC, as a VIN.

The next hurdle for the community was to consider how to enable each veterinarian to participate without feeling any difference in their capacity to operate the device. Adhering to the design concept of "ease of use by any veterinarian" applied to overcome the problem of information literacy, the community developed a simple, high-quality, low-cost videophone in September 1997, and customized it for the VIN.

The VIN that was developed connects to an ISDN and, by transmitting and receiving voice and video images, enables face-to-face discussion while providing a range of imagery (X-ray, CT, MRI, and pathological photography). Since the research workers all have the same (so completely interchangeable) equipment installed, the ability to exchange information among all research workers is an important feature. The functions of the videophone itself can be greatly extended by connecting peripheral devices such as an external camera and external monitor. It was possible to keep costs to around 400,000 yen for a complete set. Moreover, a system that can be accessed by veterinarians throughout the country has been developed for successive storing of various types of image data (such as cases and instances transmitted in real time) in VOD. In this way, comparatively low-cost technology has been established that can transmit high-definition images of various case types to facilitate remote diagnosis and can also be used in the field of veterinary medicine.

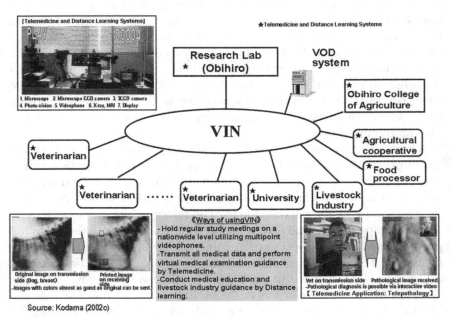

Figure 5.3.   Telemedicine and Distance Learning Systems That Were Developed and Their Manner of Use

Thus a VIN, using videophones, was developed with real-time functions (interactive related), and VOD was developed with non-real-time functions (accumulation related) (see Figure 5.3).

Innovation in Community Knowledge through VIN

One of the basic requirements for the continued development of a created community is innovation of community knowledge in terms of information, knowledge, and expertise, which will follow the steps of sharing, inspiration through contact, creation, and accumulation as the community knowledge creating cycle described in Figure 2.2 in Chapter 2 (Kodama, 2000) (see Figure 5.4). This process may be described, in simplified terms, as follows.

The first process (sharing) involves sufficient dialogue resulting in understanding between parties concerned. This understanding regards the visions and objectives pursued by different organizations in order to understand and share each other's knowledge. The second process (inspiration through contact) involves inspiring and multiplying within the circle of organizations various aspects of community knowledge to identify problems, challenges, and solutions, so that the vision and concept can be realized on the basis of the community knowledge shared by different organizations concerned. The third process (creation) involves creating new community knowledge on the basis of the community knowledge

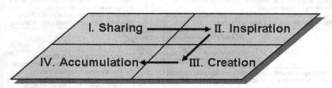

I. Sharing
    Understanding and sharing existing knowledge held in the
    community
II. Inspiration
    Propagating knowledge through inspiration related to existing
    knowledge
III. Creation
    Creating of new knowledge
IV. Accumulation
    Storing diverse new knowledge born in the process of
    inspiration, propagation, and creation

Source: Kodama (2002c)

Figure 5.4.   Process of Innovating Community Knowledge

inspired and multiplied within the circle of organizations concerned. The fourth process (accumulation) involves methodically and by organizational effort accumulating within the community, as valued expertise, the various aspects of the community knowledge harvested through the processes of sharing, inspiration, and creation.

The Strategic Community Startup Phase (see Figure 5.5)

The issue of the strategic community at the time of startup was to enlarge the community of veterinarians by disseminating and expanding

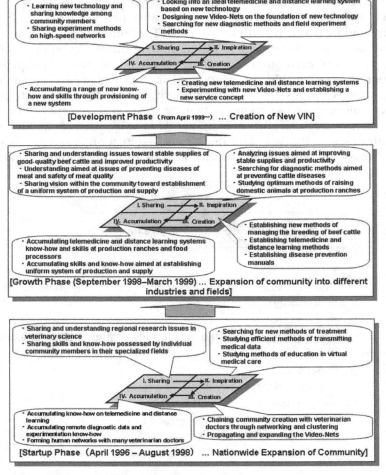

Figure 5.5. Community Knowledge-Creating Cycle

the VIN to veterinarians and universities throughout Japan. Specifically, in the first step, VINs were brought to core veterinarians (including the clinics of practicing veterinarians) in each area of the 48 prefectures nationwide, and a virtual community was formed via videophones over a network among these locations and a research institute in Obihiro City, Hokkaido. In the second step, the core locations in each of the 48 areas were made into regional centers, and by linking other new veterinarians and universities in each area, a cluster of communities was formed (see Figure 5.6).

Through the formation of these virtual communities, the veterinarians in the various regions used the VINs to engage in and spiral up the innovation process of sharing, inspiring, creating, and accumulating community knowledge. The veterinarians at the various locations became deeply familiar with each other's specialized research topics (or topics in which they lacked expertise) and with the special regional issues for study that were already accumulated in the non-real-time function of VOD, enabling them to understand the community knowledge and to share it with one another. And at fixed periods, they held a virtual "nationwide distance multipoint teleconferencing study forum for veterinarians," linking the facilities at various locations via videophones (Hirose, 2000; Tokachi Mainichi News, 1999a). The thorough dialog and discussion that occurred at these virtual

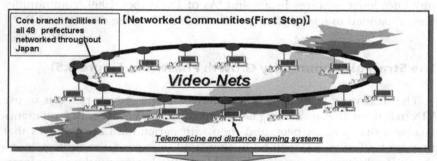

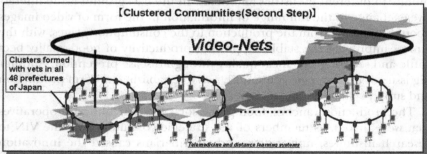

Source: Kodama (2002c)

Figure 5.6.   Nationwide Expansion of Communities

conferences inspired existing knowledge that was shared among the participants and created new methods of medical practice known as telemedicine, a new community knowledge.

Specifically, these activities fostered a medical practice conducting diagnoses and giving medical guidance in real time among veterinarians in distant locations. For example, X-ray, CT, MRI, microscope, and other photos were transmitted in real time to veterinarians specializing in teleradiology and telephasology, and appropriate diagnoses and treatments were performed virtually. Data on these cases and the results of the treatment were then accumulated in storage for VOD. A VIN was also actively utilized for distance learning. Through the VOD, the veterinarians all brought their knowledge on leading-edge medical topics in advance, they each learned, understood, and shared, and aimed to actively inspire and create community knowledge by real-time interactive communication through videophones. The new community knowledge that was obtained through these distance learning methods was efficiently accumulated in VDO, and this accumulated community knowledge was shared also with veterinarians who were unable to participate in distance learning.

In this way, the virtual methods of telemedicine and distance learning utilizing VIN are operated efficiently in the strategic community, and the innovation cycle of sharing, inspiring, creating, and accumulating community knowledge revolves in a spiral. As of December 1999, communities have expanded to several hundred locations nationwide due to this VIN.[9]

The Strategic Community Growth Phase (see Figure 5.5)

The next issue of strategic communities in the growth phase is to use VIN to link ranches producing large animals such as cattle with food manufacturers that process beef, and then with agricultural cooperatives that manage beef sales. VIN will be further used to expand the strategic community to different industries and fields. In the background is the interactive exchange of the full range of information in the form of video images about beef cattle, from the production to the consumption stages, with the aim of improving the stable supply and productivity of good-quality beef cattle and securing safe meat quality through disease prevention. Another big issue of the strategic community was to establish a uniform production and supply system at an early time.

The production ranches, food processors, and agricultural cooperatives that were linked, as members of the strategic community, via the VIN to research institutes, universities, and veterinarians caused the innovation process to develop as a spiral through dialog and collaboration among themselves. Food processors dismember and inspect the beef cattle transported to them from the production ranches, collect video data of the meat and organs from the veterinarian viewpoint (naked-eye observations and the pathological structure), and accumulate this data in VOD. Specialized veterinarians at research institutes, universities, and other organizations retrieve this video data from VOD, the data is shared among the veterinarians, and the veterinarians are able to thoroughly determine the current quality of the meat. By discussing problem areas and issues concerning meat quality via videophone, they inspire existing community knowledge and create new community knowledge in terms of solutions to be taken. Specifically, this new community knowledge becomes a manual on taking a fresh look at methods of raising and managing beef cattle and preventing disease. In this manner, production ranches and agricultural cooperatives can receive instant guidance from specialists via videophone, and take immediate action to improve the health care of beef cattle and the quality of meat.

This compares the practice of telemedicine and distance learning within the strategic community with the practical details during the startup phase, and fosters further development. The diverse expertise that results is efficiently accumulated in VOD so that community knowledge can be shared within the strategic community. In this way, a uniform system for production and supply is established in the growth phase and a regional cooperative body concept promoted. This means of utilizing video nets became Japan's first incubation in the field of veterinary science.

The Strategic Community Development Phase (see Figure 5.5)

The development phase of the strategic community comprised the challenging step of developing new technologies and services. While installing VINs developed by the community at veterinarians nationwide and gradually expanding communities in different industries and fields, the community also endeavored to develop new systems that incorporated additional new technologies. The VIN that had been planned for gradual, nationwide installation in the growth phase was based on ISDN communications networks. This was because ISDN offered the most stable line quality at that time, and provided the best solution for realizing interactive video communication. Over the next few years (in the 21st century), however, the community felt that significant developments in broadband network technology had made it important to develop new telemedicine and distance learning systems based on these communications infrastructures. The community then engaged in developing new VIN-based telemedicine and distance learning systems that ran on broadband networks, and began to build new VINs.[10]

As a concrete field trial utilizing the new telemedicine and distance learning, the community linked the research institute in Obihiro and the food processor in the Kansai region over a broadband IP-based network, and performed medical examinations on large meat animals, such as beef cattle, from a distance. Before these large animals were processed into meat, specialized veterinarians could use the virtual format to examine the animals for mad cow disease, O-157, and other communicable diseases.

The strategic community was able to use the broadband network to make telemedicine and distance learning practicable. While using this new system and aiming to establish it in fields that could be termed "the science of distance x-ray" and "distance pathology," they also actively engaged in research aimed at establishing a new educational system among different industries and fields based on distance learning. Through trials such as these, the strategic community promoted the innovation process of the community knowledge cresting cycle.

Through the above phases, the strategic community itself became an object of study aimed at disseminating VIN-based telemedicine and distance learning in the area of veterinary medicine including different industries and fields, and the strategic community worked at innovation that resulted from the spiraling-up of the community knowledge creating cycle.

The Community Competence Sophistication Process

Through the use of VIN, the strategic community members further advanced their own core competencies and those of the overall community while fostering the spiraling innovation of the community knowledge creating cycle.

To sustain continuous innovation by a community, the strengthening of core competence within the community (referred to as "community competence" in this book) becomes an important factor (Kodama, 2001a, 2002c). Community competence is high-quality, practical knowledge as inherent thought and behavior patterns embedded among community members (Kodama, 2007a, 2007b). Community competence comprises the various resources in the community that constitute information, knowledge, expertise, and conceptualization, and the competencies possessed by individuals correspond to those possessed by the community, that is, its member groups and the overall organization. Specifically, community competence is the core competencies possessed by the overall community for the purpose of incorporating resources within and outside the organization. The aim of this is to vigorously move forward with a variety of businesses as part of a creative new enterprise development effort with its roots in business reform by middle and lower level employees (the constituent members of the community). An important point is that the knowledge, expertise, and ideas of the individuals, groups, and the overall organization be stored, and then continually created and modified.

As shown in Figure 5.7, while innovation (the startup, growth, and development phases) realized through the spiraling-up of the community knowledge creating cycle creates new values, the process of community competence sophistication (sharing to creation to renewal) also becomes indispensable for ongoing strategic community formation.

In the sharing mode process of community competence, the core competencies of individual veterinarians are understood within the community. In this step, the mutual core competency is inspired into the merging phase. For example, medical skills, diagnostic know-how, and other core competencies possessed by individual veterinarians are shared among all the vets, and the inspiring and merging of these core competencies gives birth to a community competence that creates the new virtual methods of VIN-based telemedicine and distance learning.

In the creation mode process of community competence, the shared core competencies are merged to produce a new creation in the form of a

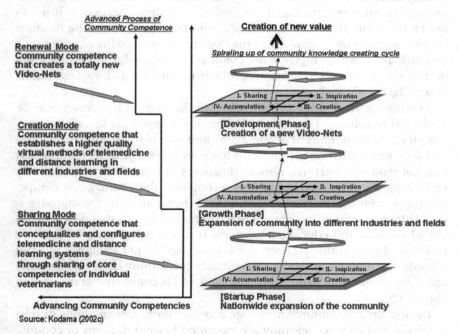

Source: Kodama (2002c)

Figure 5.7.   Advanced Process of Community Competence

product or service. In this step, development through trial and error results in new virtual methods of VIN-based telemedicine and distance learning in the field of veterinarian science, and the community of veterinarians expands outside the community into different industries and fields, such as production ranches and food processors. This in turn leads to the birth of core competencies within the community, which creates virtual methods of telemedicine and distance learning of still higher quality as a result of further attempts at improvement.

In the renewal mode process, the community competence that was advanced from the sharing mode to the creation mode is developed further. The new telemedicine and distance learning system that utilizes broadband networks is the fruit of community competence innovation. It represents the birth of a new innovation and a complete departure from previous systems.

Interactive Linkage between Community Knowledge and Community Competence

One more important point here is how crucial the two elements of community knowledge and community competence are to the strategic community as strategic management assets, and how necessary it is to interactively link these two assets if a continual innovation of the business community is to be achieved. In other words, we are looking at an interactive linkage in which high-quality community knowledge promotes community competence sophistication, which in turn feeds into community knowledge of still higher quality.

The interactive linkage as referred to in the present case study may be explained as follows (Figure 5.8). During the community knowledge startup phase, new methods of telemedicine and distance learning are created and accumulated as community knowledge by utilizing the VIN to transmit medical data and perform remote diagnoses and medical education. By expanding this VIN to veterinarians throughout the country, the competencies of many community veterinarians in the community are shared, giving birth to a community competence (in sharing mode) that disseminates virtual medical and educational systems (Process A in Figure 5.8). When this shared community competence is harnessed to put telemedicine and distance learning into practice, high-quality community knowledge, closely adhering to customers and sites, is created and accumulated (Process B in Figure 5.8).

During the community knowledge growth phase, new methods of telemedicine and distance learning are extended beyond the area of community veterinarians into the different industries and fields of production

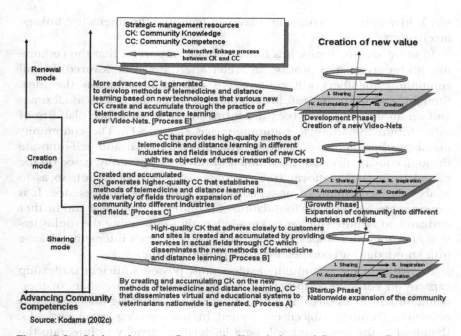

Source: Kodama (2002c)

Figure 5.8.   Linkage between Community Knowledge and Community Competence

ranches and food processors, and community knowledge of higher quality is created and accumulated. As a result, VIN-based telemedicine and distance learning takes root in a broad field, and a higher quality community competence (in creation mode) is born in the community (Process C in Figure 5.8). Aiming for more engaging innovations, the community, with its more advanced competence, inspires the creation of new community knowledge with the purpose of realizing telemedicine and distance learning that utilizes advanced technology that did not previously exist (Process D in Figure 5.8).

During the development phase of community knowledge, new and advanced telemedicine and distance learning that supports broadband networks is realized, and new discoveries and awareness that did not exist in previous VIN-based telemedicine and distance learning systems are inspired in the community through practical applications of this system. These are then created and accumulated as new community knowledge. The growth in skills, know-how, and other assets relating to VIN-based telemedicine and distance learning born from these new technologies is further enhanced, giving birth to more advanced community competencies (in renewal mode) within the community (Process E in Figure 5.8).

This interactive linkage can be attained through the innovative leadership of the community leader, the important point being that the community leader must be a conscious player in handling the interactive linkage mechanism.

In more specific terms, it is fundamentally important that the community leader create and provide an arena where it can be assured that all community members will continue to uphold in common the idea, thought, and spirit that will inspire them to create new value at all times and on an ongoing basis (i.e., the formation of a resonance platform of value within the strategic community: see Figure 5.1). The community members then endeavor, individually, to self-improve and self-innovate through constructive dialog promoted within the community, based on the value-harmonized platform (the resonance platform of value), so as to enable the acquisition of new knowledge, and hence competence. It is important to build on the foundation of acquired competence and further endeavor to acquire still newer knowledge, thus enhancing the sophistication of one's own competence and so promoting one's interactive linkage with knowledge and competence on a personal plane.

Accordingly, the community leaders must possess sufficient leadership capacity to continue empowering and motivating the members of their communities on an ongoing basis, with an eye to attaining the community objectives (business objectives) rooted in harmonizing value outlooks. Required at the same time will be a leadership capacity to integrate the knowledge of the individual members of the community with the interactive linkage process of competence on the collective and organized level, and so establish such interactive linkage processes as a community-wide management system on firm ground.

ICT Dynamic Capability by Strategic Community

As explained above, the strategic community creates diverse new knowledge in each of the three phases. Then in each phase, the strategic community actors adjust the three elements of their acquired ICT dynamic capability, namely, ICT application capability, context architect capability, and boundaries consolidation capability (see Figure 5.9). Considered from the viewpoint of the strategy-making process, the actors' implementation of strategy developed through trial and error (emergent strategy), simultaneously incorporating advancing technology and the needs of the workplace, was a common pattern in all phases.

In the startup phase, the actors' "productive interpretation" of innovation in the veterinary science domain and the "creative realization" of the possibilities for remote diagnosis using ICT became a trigger for actors enhancing "ICT application capability" aimed at the challenges of developing ICT. Moreover, the new contexts of the need for remote diagnoses and raising vet-

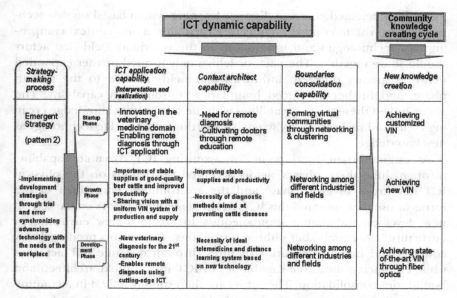

Figure 5.9. ICT Dynamic Capability and New Knowledge Creation

erinarians' training levels still further through remote education raised the actors' context architect capability. The actors deliberately formed virtual communities through network and clustering in response to new contexts, and raised their boundaries consolidation capability. Then the actors were linked by these three capabilities and the community knowledge creating cycle, and achieved a customized VIN system as new knowledge.

In the growth phase, two triggers emerged to enhance actors' ICT capabilities of use. One was the productive interpretation of the importance to actors of stable supplies of good-quality beef cattle and improved productivity; the other was creative realization from the viewpoint of a vision with a uniform VIN system of production and supply. Moreover, the focus on improving stable supplies and productivity and the need for diagnostic methods aimed at preventing cattle disease produced new context architect capability among actors. Actors deliberately formed networks among fields and industries and enhanced boundaries consolidation capabilities based on new contexts. Then they realized a new VIN system as new knowledge linking these three capabilities and the community knowledge creating cycle.

In the development phase, the "productive interpretation" of how actors should carry out new veterinarian diagnoses in the 21st century and "creative realization" from the viewpoint of pursuing the possibility of remote diagnosis through cutting-edge ICT development further enhanced the actors' ICT use capabilities. Moreover, the focus on the need for an ideal telemedicine and distance learning system based on new technology gave rise to context architecture capability, a new context comprising new technologies and innovations in the veterinary field that actors should share together. Then actors deliberately formed further expanded networks among different industries and fields relative to the growth phase, and further enhanced boundaries consolidation capability. The actors linked these three capabilities with the community knowledge creating cycle, and realized a VIN system supporting cutting-edge technology as new knowledge.

As an important focal point for acquiring ICT dynamic capability through these three phases, actors accepted restrictions on the physical ICT developed in each phase and the environment (structure) of the forms of use that activated this ICT. At the same time, actors embarked on new development aiming spontaneously to create a new environment (structure) once again, but without restrictions. Then actors produced new contexts and human network structures through dynamic ICT activities while enhancing the three capabilities of ICT use, context architect, and boundaries consolidation. The actors are then spirally linked in a community knowledge creating cycle through these three capabilities, and go on to create new knowledge. After that, they go on to acquire new ICT dynamic capability (see Figure 5.10).

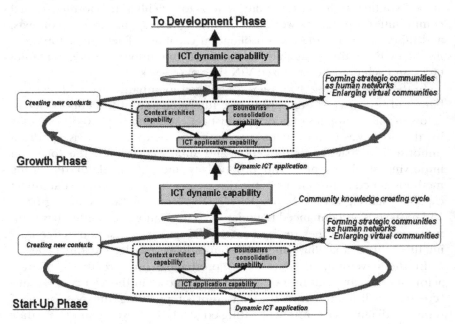

Figure 5.10. Dynamic Loop of ICT Activity

The Superiority of VIN

Interviews and questionnaire surveys were conducted among 63 members of this strategic community concerning VIN from the viewpoint of improvements in business process and quality. The content of the survey focused mainly on the superiority of VIN compared with other existing text-based information systems such as email, groupware, and text databases (see Figure 5.11).

Based on these results, we found that VIN is superior for promoting strategic community management in the community setting in the following ways. VIN is superior in bringing efficiency to community business processes by expediting the decision making of community leaders in order to promote strategic community business, and by fostering the high-quality communication and collaboration within and among communities that is embodied by interaction among individuals. VIN is also superior in sharing, inspiring, creating, and accumulating community knowledge, and then advancing community competencies.

At present, many companies and heterogeneous organizations are installing databases and groupware designed to handle documents and other text information, and are promoting knowledge management within their organization or between organizations. According to advanced users

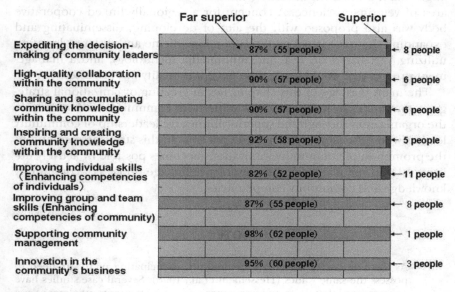

Source: Kodama (2002c)

Figure 5.11.   Superiority of VIN Over Groupware and Email

in Japan, Europe, and the United States, however, it is becoming apparent that these text-based systems alone do not allow them to sufficiently strengthen communication or expedite decision making, whether inside or outside the organization (Kodama, 1999a, 1999b ; Nikkei Information Strategy, 1998).

In short, issues related to the one-directional, non-real-time nature of contact and consultation by email, groupware, and other systems are being raised as the greatest obstacle against making important judgments and decisions quickly. This is because the most important element in organizational communication is the empathy and solidarity between individuals that comes with face-to-face communication.

In this respect, the VIN advocated in this chapter completely sweeps away the "isolation" in which individuals' thoughts and intentions cannot be sufficiently communicated through email, groupware, and the like. VIN becomes the ultimate tool of communication that leads companies and organizations toward innovation.

Conclusion

In this case study, the management style of "strategic community creation" among different industries has been actively incorporated into the area of veterinary science. A concept for a regionally linked cooperative body was also proposed with the aim of developing, disseminating and expanding the new virtual methods of telemedicine and distance learning utilizing VIN based on ICT and multimedia technology, and a management process that would bring this concept to reality was discussed.

The main message of this chapter lies in creating organizational strategic communities by resonating new values with community leaders outside the organization through the superior innovative leadership of community leaders who possess new visions and concepts. In this strategic community, the promotion of community innovations becomes possible by actively utilizing VIN to spirally develop an interactive linkage process of community knowledge and community competencies.

NOTES

1. As a condition for forming a community, participating members need to possess the same values (Hesselbein et al., 1998). Several case studies have also reported that the values resonance process is an important element as a condition for the formation of strategic community creation (Kodama, 1999b).

2. The community knowledge innovation process refers to spiraling development of the Sharing to Inspiration to Creation to Accumulation process whereby community members engage in dialog and collaboration on a variety of information and knowledge within the community (Kodama, 2000, 2004). In terms of general corporate activities, this community knowledge spiraling process involves repeated improvements of quality in the development and marketing of products and services more attuned to customers' needs, which in turn raise the quality of community knowledge with each repetition.

3. Community competence consists of the various resources in the community that make it up, "skill, experience, expertise, and conceptualization, etc.," and the competencies possessed by individuals corresponding to the competencies possessed by the community (i.e., its member groups and the overall organization). Specifically, community competence represents the core competencies possessed by the overall community for the purpose of incorporating resources within and outside the organization including customers, in order to vigorously move forward with a variety of businesses as part of a creative new enterprise development effort with its roots in business reform by middle and lower level employees who are the constituent members of the community. An important point is that the competence of individuals, groups, and the overall organization be shared, and then continually created and renewed (Kodama, 1999b). On the other hand, concepts such as new product development capability (Prahalad & Hammel, 1990), capability to promote business processes (Stalk, Evans, & Schulman, 1992), and the capabilities of organization members (Grant, 1991) have already been discussed in previous research reports, however in this book, community competence refers to the sharing and merging of competencies, which are the capabilities and competencies of many types of groups and organizations, and to the creation and renewal of new competencies that the strategic community aims for in achieving its mission and goals.

4. This means that the relationship between the elements of community knowledge and community competence inside a strategic community must possess a mutually interactive connection at the individual and community levels. In other words, the acquisition of new community knowledge enhances the quality of community competencies, and the enhanced community competencies induce and inspire the acquisition of additional new community knowledge. This is known as the interactive linkage of community knowledge and community competencies, and we describe this structure in detail through this case study.

5. The importance of the process in which dialog and collaboration between individuals give birth to new knowledge has also been discussed in previous research (Nonaka & Takeuchi, 1995; Watkins, 1993).

6. In recent years, veterinary medicine has been in the process of advancing toward the prevention of various diseases with organ transplants, and at gene level, explaining the phenomenon of life. These days, when veterinary medical treatment is being questioned on the basis of life ethics, from a wide-ranging viewpoint that integrates medicine, engineering, and science, "the education of talented personnel and the establishment of research and development having the characteristics of veterinary medicine in the chemical view of data related to animal medicine" is essential for the mastery of veterinary medicine. The place where veterinary medicine differs from

other natural sciences is that, having discovered the meaning and value of the existence of all animals, it helps them to survive. With respect to the existence of industrial animals and companion animals, because of the existence of social meanings and values in each case, clinical veterinarians carry the burden of fulfilling that duty everyday with a great sense of mission (interview with Dr. Hirose).

7. Hirose is the Director of the Animal Medical Information and Science Development Research Institute and Professor Emeritus at Obihiro University of Agriculture and Veterinary Medicine. He specializes in clinical veterinary radiology. Using Japan's first imaging diagnostic terminology, he is a world pioneer in contributing to the development of imaging diagnosis by developing and introducing large-animal X-ray diagnostic vehicles and industrial-animal general imaging diagnostic vehicles, which he continues to actively bring into the farmyard. Dr. Hirose was awarded the Japan Veterinary Association President's prize in 1972. He is a director of the International Veterinary Radiology Association and sits on the review board of the Veterinary Radiology and Ultrasound Journal. Dr. Hirose is the chair of the organizing committee for the 12th International Veterinary Radiology Association (IVRA) that was held in Obihiro, Hokkaido, in August 2000.

8. Interview with Dr. Hirose.

9. "Image transfer technology for remote diagnosis and treatment technology," a series of research activities for this strategic community, has been adopted by the "Fiscal 1998 new project creative research and development system" of the Telecommunications Advancement Organization of Japan (TAO, a corporation approved by the Ministry of Posts and Telecommunications that is eligible to receive funds from the Japanese government for research and development) (Tokachi Mainichi News, 1999b). This result of community innovation was acknowledged by government organizations and has further heightened the motivation of community members toward greater expansion of the community. TAO is a corporation, approved by the Ministry of Posts and Telecommunications, that provides various types of assistance in respect to research and development and telecommunications advancement projects in the field of data communication.

10. In 1998, TAO constructed a super-fast optical fiber communications network for research and development (commonly known as the gigabit network) to be compatible with the next-generation Internet and intended for use by academic research organizations all over Japan. TAO launched an appeal to the public, including research organizations throughout Japan, in connection with "using the gigabit network for research and development." In this connection, "Image transmission experiments for a telemedicine system in the areas of animal care treatment and stockbreeding," proposed by the strategic community, was selected by competitive tender in 1999 (Tokachi Mainichi News, 1999c).

CHAPTER 6

ENABLING EMERGENT AND DELIBERATE ICT STRATEGIES

In this chapter, I consider the thoughts and behavior of actors engaged in justifying new knowledge in-house following its creation from a bottom-up approach through emergent strategy, and look at initiatives to enhance in-house productivity and customer services as a company-wide, deliberate strategy. This chapter also points out that fully utilizing VIN supports managerial excellence while enabling the formulation of new business models for creating customer value. Aggressive ICT investment by top management to build a VIN (a support tool for this future network strategy) will permit business innovation based on strengthened competitiveness and enhanced customer service.

This chapter looks at a real example of communities within companies, among companies, and between companies and their customers with the case of a company currently making active use of VIN. The in-depth case study of a manufacturing company demonstrates the value of VIN as a network strategy support tool that bolsters community management in companies.

New Knowledge Creation Through ICT Dynamic Capability, pages 125–139
Copyright © 2008 by Information Age Publishing

CUSTOMER VALUE CREATION MANAGEMENT
(IBIZA, INC.)[1]

The Starting Point for Customer Value Creation

IBIZA, Inc.,[2] a Japanese manufacturer of high-quality leather bags for women, is one of Japan's industry leaders, winning the Japan Quality Award in 1998[3] (Nihon Keizai Shimbun, 1998). IBIZA has established a production system incorporating every stage in the bag's life cycle, from careful selection of leather to manufacture, wholesaling, retailing, and after-sales service. By establishing atypical sales channels that bypass wholesalers and allow it to deal directly with retail outlets, and by designing, manufacturing, selling, and repairing its own bags, IBIZA has developed a highly profitable management style that has seen a 20% increase in nominal profits over the past 10 years (see Figure 6.1).

An unusual feature of the company pointed out by observers is that it manufactures and sells its original-design "warm, all-natural, handmade bags" in various styles and small lots, with a lifetime guarantee. In consideration of company management in the customer value creation model, and to further win over its customers, IBIZA offers a variety of substantial after-sale services.

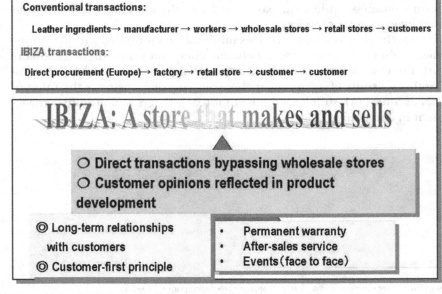

Figure 6.1. IBIZA's Vertically Integrated Management

Regarding these after-sales services in particular, IBIZA can be seen as a new kind of business that wins intense loyalty from customers. IBIZA not only offers informational services that include sending more than 50,000 direct mailings and copies of *IBIZA Magazine* to fans of the IBIZA brand each year, but also makes an impression on each customer who purchases a bag by sending a personal letter of thanks from the company president.

Past Issues

Making products that utilize nature in its original state is the concept and value behind IBIZA. In practical terms this means using, in its original form, animal leather that has been affected by variations in the natural environment. A piece of leather from an animal's abdomen, for example, differs greatly from one from its back. That is why, despite being a standardized article, each IBIZA product is a little different from the next.

The company has been searching, through trial and error, for a way to communicate with its customers and salespeople about products that is faithful to each individual article. A similar situation exists with repairs. IBIZA products come with a lifetime guarantee. The company used to take repair requests by telephone and fax, but customers found it difficult to get across the subtle nuances of their requests. So it was important to contrive a way for repair staff members to communicate with distant customers as though they were holding the product in question in their hands.

As a company based on direct sales that bypasses wholesalers, IBIZA was pondering a method of direct communication with its customers and dealers that CEO Mr. Yoshida himself could oversee. Under the constraints of time and distance, voice and text-only systems such as telephone, fax, email, and groupware somehow could not solve these problems. The company needed to create a "take a look and you'll see what I mean" environment with the feel of television or video.

PRODUCTIVE INTERPRETATION AND CREATIVE REALIZATION

At that time, IBIZA's relationship to ICT mainly involved management of the legacy customer data processing system, and almost no resources were directed to the theme of corporate innovation through implementing advanced ICT. But Mr. Noriyasu Koguchi, head of the management planning division and who was also in charge of the in-house information system, retained a strong sense of the past issue mentioned above. Mr.

Koguchi sought out a future ICT strategy of IBIZA's own while supervising the management planning business.

But rather than IBIZA, as a small or medium-sized company, arranging for a dedicated in-house ICT supervisor, Mr. Koguchi himself took charge of planning for the entire ICT operation. At that time, Mr. Koguchi expressed interest in a desktop-type video-conferencing system for VIN video terminals. This system operated on Windows 95, and possessed functions enabling the sharing of bidirectional multimedia communication and data using ISDN. Mr. Koguchi's interest in this VIN led to the "productive interpretation" of enabling ICT business process innovations and enhancing customer services. The challenge of experimenting with the application of ICT through staff at the workplace organization, centered on Mr. Koguchi, also led to a "creative realization" among members.

Mr. Koguchi was soon holding repeated discussions with the person in charge of sales at a workplace organization (local branch) of NTT, Japan's biggest communications carrier. Before purchasing and deploying the video terminals, IBIZA had to quickly confirm their effectiveness through collaboration with the communications carrier. Using the latest image compression technology, experiments began as to whether a bag's faulty areas and such points as faded color could be accurately conveyed. Trials were conducted with both still and moving image transmission under a variety of conditions with regard to camera specifications, lighting, and the background color and microphone placement for the photographic subject. Bag designers, product planners, production workers, people in charge of repairs, shop sales staff, and others participated in the experiment, and provided comments via the video terminals.

About a month later, through a trial-and-error process, those involved in the experiments came to form a shared recognition of the fact that it was an effective system for the VIN to verify faded bag color or fault areas remotely via camera image transmission. Prior to and during the experiments, those in charge had voiced a variety of opinions, and came to form a common perspective through dialog among employees taking part in the trials, despite the existence of equivocal contexts among employees' boundaries communication. This linked to a positive, standardized result from implementing emergent strategy amid the equivocal context relating to the IBIZA workplace organization, and a common perspective was formed regarding VIN applications among employees.

The final decision on IBIZA's ICT installation was implemented by CEO Shigeru Yoshida. Because of this, the VIM trial team, centered on Mr. Koguchi, decided to make a presentation to Mr. Yoshida on the theme of enhancing business process reform and customer services that apply VIN tools. Mr. Yoshida himself did not have the technological knowledge to understand the state-of-the-art VIN, and decided to introduce it from the standpoint of how it contributes to enhancing customer service. Then the experimental team led by Mr. Koguchi began to share the common perspectives regarding ICT strategy through dialog with Mr. Yoshida.

In this way, Mr. Yoshida handed down the decision to introduce VIN systematically throughout the entire IBIZA company. IBIZA also started up an in-house project to implement ICT strategy initiatives. Next, this project was settled on and implemented as a result of a deliberate strategy to install VIN full-scale in the next stage. The behavior of these organization members in regard to IBIZA involved a management shift of implementation from an emergent strategy based on trial and error to a deliberate strategy.

ICT INVESTMENT FOR THE CUSTOMER

So the company undertook a large-scale ICT investment project to construct a VIN focused mainly on connecting IBIZA (its headquarters and factories) with dealers throughout the country using video-conferencing systems, videophones, or similar devices.

The combination of a company with a slogan like, "warm, all-natural, handmade bags," and multimedia video information distribution may seem paradoxical. From the standpoint of customer value creation and management innovation, however, it indicates an important element in IBIZA's unique brand of company management.

The company's reason for taking the plunge into ICT investment was not simply to bring efficiency to its business operations using the latest multimedia, but also to innovate, share, and develop the concept and value of each individual product, or "composition" (the company goes so far as to call each individual product a composition) with employees, customers, and dealers using video information. Yoshida's customer-oriented ideas and superior leadership were the basis of this investment in "ICT for the customer."

VIN SUPPORTING CUSTOMER VALUE CREATION MODEL BUSINESSES (TYPE 2 AND TYPE 3 COMMUNITIES APPLICATION IN FIGURE 4.1)

By the end of January 1996, IBIZA had built a network on its own premises and those of its dealers, and was using it to send product information, take orders, and carry out maintenance business. In this way, the community between IBIZA and its dealers and the community between the company and its customers were put to use. Specifically, the VIN had two broad objectives. The first was to win orders by sending information on newly created products to dealers throughout the country without delay; and the second was to take orders and check inventory quickly and accurately.

In the past, methods for announcing new products were limited to exhibitions, private viewings, and the semiannual publication of a catalog, but by introducing a VIN into a segment of the dealers in Japan, IBIZA was able to easily offer those shops detailed information on new products without a moment's delay, and to obtain purchase orders on the spot using bidirectional video communication. Using the VIN to make product announcements eliminates the decorating and traveling expenses incurred in preparing venues for conventional product announcement gatherings. Because each piece of the leather used for making bags looks different, dealers find it desirable to inspect products and get detailed inventory information when placing orders. This is another area where the VIN is demonstrating its power. This application of the VIN in the community of IBIZA and its dealers is garnering attention as a new method of marketing through one-on-one interaction.

Application of the VIN to properly respond to customers' product maintenance needs is another important business process. IBIZA products carry a lifetime warranty. Connecting IBIZA factories with dealers by means of a VIN allows repair staff members to directly observe images of the repair locations, check work, and respond to inquiries, permitting an accurate response to all the nuances of customer requests with no misunderstandings. There have also been cases of product planners or factory supervisors giving direct product demonstrations to customers at a dealer's shop over the VIN, with the demo leading to a new order. This can be seen as a new form of community management that includes customers as well as IBIZA and dealers. It consists of virtual, real-time contact between the IBIZA corporate organization and customers over the VIN, and provides an example of a business in the customer value creation model eliciting trust and security from a customer.

With a community such as this established between a company and its dealers, major assets are the sharing, inspiring, creating, and accumulating of new community knowledge and the ability to earn the sympathy and trust of customers through interaction consisting of communication and collaboration.

CONTINUAL SHARING, INSPIRING, AND CREATING COMMUNITY KNOWLEDGE (TYPE 1 COMMUNITY APPLICATION IN FIGURE 4.1)

As the first step in configuring a VIN, IBIZA promoted community management with its customers and dealers. As the second step, to enhance the shar-

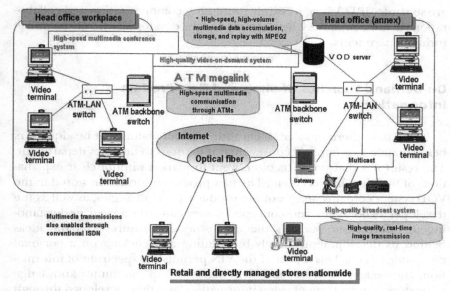

Figure 6.2.   IBIZA's VIN Network Organization

ing, inspiring, creation, and accumulation of community knowledge within the company, it built a cutting-edge VIN on a high-speed, wideband, fiber-optic ATM network using product images to function as an order placement and reception system (Nikkei Sangyo Shimbun, 1998) (see Figure 6.2).

While promoting communication and collaboration among individuals in the community between its headquarters and factories, IBIZA also aims to store community knowledge such as new product planning and shop expertise in the form of video information, share this knowledge, and engage in new knowledge creation and innovation.

Communication and Collaboration Using High-Quality Video Information

To respond quickly and accurately to rapidly changing market needs and various customer demands, businesses must raise the level of their customer service through such means as shortening the new product development period. For these purposes as they relate to development, there is a need to carry out design, development, and manufacturing-related interaction more quickly, accurately, and frequently. High-quality video information used over a VIN offers an environment where a consultation appears to be held in a single location, even when the headquarters and factories are separate. IBIZA is aggressively promoting communication and collaboration in the community between its design, development, and manufacturing departments.

On-Demand Searches of Shop and New Product Information

In the past, there was a problem with product planners at headquarters being unable to accurately convey their intentions to the sales department. The result was that products often went to market without clear explanations of the ideas behind them. The new product information stored in the VOD system contains images of the products from all angles, as well as the developer's comments and concepts. A user can gain a clear understanding of information, such as the new product's features and the ideas behind its development, merely by pointing and clicking on a personal computer screen. This aspect of the VIN permits the spectrum of information, knowledge, expertise, and ideas that make up community knowledge to be stored in the form of video information, and then developed through further creation and innovation (see Figure 6.3).

Salespeople used to make the rounds of their assigned shops, and then include in their daily sales reports detailed written information on product

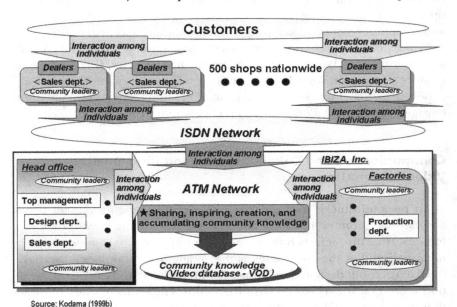

Source: Kodama (1999b)

Figure 6.3. Community Knowledge Creation by VIN at IBIZA, Inc.

displays and the situations in stores. But these reports were somehow unable to convey an image of the customers and bags in the shops, and when personnel changes brought a new salesperson, background information that cannot be found in figures and written reports, such as recollections of communication with customers, would disappear with the old salesperson. Today, with the newly configured VOD system, a user can take in at a glance details such as a store's layout, manager, and product display, and the system has been configured to allow headquarters to give stores detailed guidance on matters such as product display changes. This mechanism is an application of the VIN that enables the continuous contribution of novel, frontline sales styles that are in harmony with the new strategies of IBIZA's sales department. The goal is to further enhance customer service related to on-site sales.

CUSTOMER VALUE CREATION WITH THE CUSTOMER THROUGH MOBILE PHONE NETWORKS

Furthermore, IBIZA puts effort into sharing values with customers who use mobile videophones (third-generation mobile phones), with the aim of further direct marketing with the customer (see Figure 6.4). IBIZA's aim is to receive the customer's direct requests and complaints through 3G mobile phones, which are expanding worldwide, and to link them to prod-

★ **Mr. Yoshida uses a mobile videophone to talk with a customer and show a new bag from the IBIZA bag production workplace**

Source: NTT DoCoMo public relations materials

Figure 6.4.   Forging Closer Bonds with the Customer by Connecting via Mobile Phones

uct service development and enhancing customer services. Focused on Mr. Koguchi's project team, IBIZA is currently experimenting with direct dialog with the customer through mobile videophones.

> As part of customer support, our company uses mobile videophones and undertakes initiatives to transmit images of repair locations to the factory. The person in charge, even from a separate location, uses them as support tools to check in-house displays and inventories. "Any time, anyplace, with anyone," is also the motto of this company, and ideally, I would like to carry out direct marketing methodically using mobile videophones. We are in the midst of a trial service providing image data on new products and entertainment to customers' mobile videophones, aimed at members within the company and members of IBIZA's broadcast agency. Realistically, issues such as mobile phone ownership can create difficulties, but it will be fine so long as we can distribute mobile videophones to all customer members, and once we have made up our minds, quickly establish visual communication regardless of the store. (Mr. Koguchi) (see Figure 6.4)

IBIZA is enjoying a growth period as a company that sustains annual operating profits of 20%. While securing this growth is one of CEO Yoshida's basic aims, the spirit of taking on challenges to implement a range of new experiments aimed at enhancing customer services is becoming established within the company. Direct marketing using mobile phones is one of these new challenges, and it also becomes a recursion to emergent strategy from the viewpoint of the strategy-making process. Put another way, despite elements of uncertainty toward the next stage of corporate growth, new business models will be sought out using dynamic, practical ICT investment.

CREATING "ICT DYNAMIC CAPABILITY" THROUGH DIFFERENT STRATEGY-MAKING PROCESSES

IBIZA's VIN installation step (described above) comprises three basic phases (see Figure 6.5). The first phase consists of initial introduction and experimentation. The experimental phase established ICT application methods through trial and error as a result of emergent strategy. Although equivocal boundaries communication dominated at first among members of the workplace organization, learning processes through trial and error later went to create uniform perspectives of ICT strategy. The "productive interpretation" of enhancing customer services as a result of workplace organization members applying ICT, and the "creative realization" of learning through experiments applying VIN at the workplace, became the triggers enhancing "ICT application capability" aimed at the challenge of introducing ICT to organization members. The new context of how to

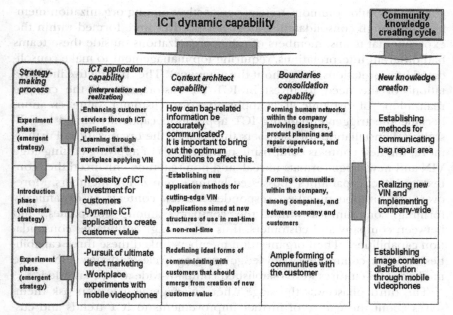

Strategy-making process	ICT application capability (interpretation and realization)	Context architect capability	Boundaries consolidation capability	New knowledge creation
Experiment phase (emergent strategy)	-Enhancing customer services through ICT application -Learning through experiment at the workplace applying VIN	How can bag-related information be accurately communicated? It is important to bring out the optimum conditions to effect this.	Forming human networks within the company involving designers, product planning and repair supervisors, and salespeople	Establishing methods for communicating bag repair area
Introduction phase (deliberate strategy)	-Necessity of ICT investment for customers -Dynamic ICT application to create customer value	-Establishing new application methods for cutting-edge VIN -Applications aimed at new structures of use in real-time & non-real-time	Forming communities within the company, among companies, and between company and customers	Realizing new VIN and implementing company-wide
Experiment phase (emergent strategy)	-Pursuit of ultimate direct marketing -Workplace experiments with mobile videophones	Redefining ideal forms of communicating with customers that should emerge from creation of new customer value	Ample forming of communities with the customer	Establishing image content distribution through mobile videophones

Figure 6.5. ICT Dynamic Capability and New Knowledge Creation

bring out the optimum conditions to accurately communicate bag-related information, moreover, enhanced the context architect capabilities possessed by the organization members. The organization members also formed human networks that dynamically involved in-house designers, product planning supervisors, repair supervisors, and salespeople to support new contexts, and enhanced boundaries consolidation capabilities. Then the organization members linked these three capabilities with the community knowledge creating cycle, and established methods for communicating bag repair areas as new knowledge using VIN tools.

The second phase was a process of justifying, company-wide, the results of implementing emergent strategy in phase one. Here, significance was later attached to the success of the emergent strategy in the workplace resulting from top management, and the deliberate systematization became important as the company's rational strategy. Resource distribution as a result of top management decision making becomes an approach for justifying best practices company-wide through emergent strategy in the workplace. As consideration of top management strategy, this involves a shift in the strategy-making process from bottom-up emergent strategy in the initial experimental phase to top-down intended or deliberate strategy.

Then top management redefines and adjusts the aggregate of the equivocal strategy perspectives in the workplaces throughout the company, and goes on to build common strategy perspectives among organization members (although consolidated strategy perspectives were formed within the experimental teams, members of other organizations outside these teams had different interpretations, requiring top management to make consolidated interpretations throughout the company). The "productive interpretation" of the need to invest in ICT for customers, and the creative realization" of dynamic ICT application for customer value creation, became the trigger enhancing ICT application capability aimed at full-scale ICT installation for members throughout the organization. Furthermore, the new contexts comprising new methods of applying cutting-edge VIN and new usage formats in real- and non-real-time enhanced the "context architect capabilities" possessed by the organization members. Moreover, organization members responded to new contexts by dynamically forming communities within the company, among companies, and between company and customers, thus enhancing "boundaries consolidation capabilities." Then organization members linked these three capabilities and the community knowledge creating cycle, and new VIN usage methods and cultures were established company-wide as new knowledge.

The third phase was the stage where management leaders ask themselves about the nature of further improvements to ICT trends and customer services in the future. What kind of ICT is familiar for IBIZA's main customers (including housewives and the aged)? The idea that it might be mobile phones rather than PCs led management leaders to new, creative thinking. For companies in the IBIZA-type growth stage, the search for new business is the emergent strategy-making process. In IBIZA's case, however, this phase does not represent a total shift from phase two's deliberate strategy to emergent strategy. While continually implementing a company-wide, deliberate strategy, it is also a structure that permits implementation of emergent strategy in-house as a trial-and-error process. Put another way, it is the coexistence of the different strategy-making processes presented in Chapter 3.

Equivocal boundaries communication among members of the workplace organization implemented emergent strategy through trial and error. The "productive interpretation" of the pursuit of ultimate direct marketing through ICT and the "creative realization" of the new trials of workplace experiments through mobile videophones became a trigger for raising the ICT application capabilities of new ICT tools in organization members. The new contexts of redefining the ideal way to communicate with customers in order to create new customer value raised the organization members' context architect capability. Moreover, organization members aimed to enrich the dynamic formation of communities with customers in response to new contexts, and raised boundaries' consolidation capabilities. Then organization members linked these three capabilities with the community knowledge creating cycle, and sought out the possibility of image content distribution from mobile videophones as new knowledge.

An important focal point for IBIZA in acquiring ICT dynamic capability is that of top and middle management dynamically managing the different strategy-making processes of emergent and deliberate strategies on a time axis. Implementing emergent strategy will enhance new ICT application capability, centered on personnel, in the specific field of after-sales service (repair and fault support); the context architect capability of the organization members (an ICT strategy that did not exist previously); and the boundary consolidation capability to implement these ICT strategies. Meanwhile, implementing deliberate strategy will enhance ICT application capability as an objective of all employees while also enhancing context architect and boundary consolidation capabilities of organization members with regard to incremental learning aimed at improving daily work duties. Put another way, the purpose of implementing deliberate strategy will be the acquisition of ICT dynamic capability for current exploitative activities on a daily basis, while the purpose of implementing emergent strategy is the acquisition of ICT dynamic capability for exploratory activities aimed at future business.

Then organization members simultaneously enhance each of the three different capabilities (ICT application, context architect, and boundary consolidation) and create new context by building human networks from ICT dynamic application. The organization members are linked spirally to the community knowledge creating cycle through interaction with real and virtual space as a result of these three capabilities, and create new knowledge. The actors go on to acquire ICT dynamic capability to maintain a competitive edge.

Supporting IBIZA's acquisition of sustained ICT dynamic capability is the starting point for strategic objectives rooted in the corporate vision and customer-first principle of customer value creation management. IBIZA is dialectically embedding a range of customer knowledge (tacit and explicit) within the company (see Kodama, 2002b), including the management environment and customer needs surrounding the company, and carrying out multilayered consolidation of communities inside and outside the company, including customers. In a society with diversifying customer lifestyles and levels of satisfaction, dialog with customers is indispensable to obtain and expand secure, repeat business with new customers.

IBIZA's knowledge management goes beyond the free use of ICT to incorporate interaction with the customer in dense real space. The points of contact between IBIZA and customers are those of dense dialog and contact with regards to a range of sustained activity (including factory inspection tours, *IBIZA Magazine*, IBIZA fairs, repair visits, exhibitions, par-

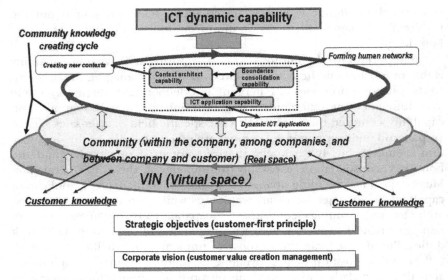

Figure 6.6. IBIZA's ICT Dynamic Capability

ties, and postcards). IBIZA organization members will go on to sustainably practice the new community knowledge creating cycle through dense dialog with customers and the community. IBIZA organization members are linked by corporate vision (customer value creation management), strategic objectives (customer-first principle), the community knowledge creating cycle linkage spanning real and virtual space, dynamic ICT application, the creation of new contexts, and the formation of human networks, and go on to acquire sustained ICT dynamic capability (see Figure 6.6).

As described above, the mechanisms at IBIZA whereby communities are formed and customer needs picked up through frequent, dense interaction with customers in real and virtual space, employees then share the knowledge gained and quickly return it to the customers' accounts for a high 20% or so of current profits.

NOTES

1. Interviews were conducted with Mr. Yoshida and the top management of IBIZA, Inc. and with the management of vendors who have been engaged in the design, configuration, and operation of IBIZA's VIN.

2. The company's product brand name is the name of the Spanish island of Ibiza, and it has earned 780,000 devoted customers throughout Japan. It has earned numerous management awards, prominent among which are the "13th Nikkan Kogyou Shimbun Excellence in Management Award,"

"the Corporate Small Business Research Center Award for the Kanto District," and "the 1995 Saitama Prefectural Sai no kuni Factory Designation."

3. The Japan Quality Award is an annual award that recognizes excellence of management quality. Award winners are companies that are managed from the viewpoints of customers, the source of business profits, and that have management framework to continuously create new values. The purposes of this award are: to innovate the whole industry; to change economic structures; and to improve the living standard in harmonization with international situations. This award system was established in December 1995. This award has seven central concepts. "Japan Quality Award Criteria" has eight categories. Applying companies submit their "management quality reports." They are evaluated in four stages by the reviewers. They must be finally qualified by the Japan Quality Award Committee.

CHAPTER 7

MANAGING STRATEGIC MANAGEMENT CYCLES

A Case Study of Sony

In this chapter, I consider the management process with regard to corporate strategy for introducing ICT as it occurs on the time axis of the strategy-making process. In Chapter 6, I discussed the shift in the strategy-making process from an emergent to a deliberate strategy, and then I looked at the process leading to a dual strategy, with deliberate strategy embedding elements of emergent strategy. This chapter also looks at the opposite process of the shift from a deliberate to an emergent strategy, and then to a recursion of deliberate strategy. The management of these different strategy-making processes indicates the importance of enhancing the company's inherent ICT application capability and producing sustained dynamic ICT capability. I have decided to analyze Sony's VIN introduction process as a case study.

SONY MARKETING'S ICT STRATEGY: A CASE STUDY

The globally famous Sony Corporation develops and sells electronic products in great numbers. Sony Corporation's company organization, Sony Marketing Japan Inc. (referred to as "Sony Marketing" below), is the company that has generally taken charge of marketing and sales for Sony prod-

New Knowledge Creation Through ICT Dynamic Capability, pages 141–150

ucts in Japan. Sony Marketing employs 2,600 staff and posted net sales of about 650 billion yen in fiscal 2004.

In 2005, Sony had the third-largest market share globally (and the second largest in Japan) in the field of videoconferencing, with its characteristic VIN system video terminals. As a leading global electronics manufacturer, Sony possesses excellent image compression technology, which is at the core of video terminal technology. Sony Marketing has installed and applies its own VIN system at its offices, and aims to exhibit a presence to rivals as a company with a single ICT model. Sony Marketing's president makes full use of his company's original products, and considers it important to persuade other companies and industries of the efficacy of introducing VIN.

At that time, the expectations that the president held for introducing VIN to Sony Marketing were threefold: to speed up decision making in-house, reduce costs (including business trip expenses), and stimulate communication in-house. The president ordered the staff of the head office organization's general affairs department to implement policies for plans to introduce what were then cutting-edge VIN systems (IP-based videoconferencing systems and multipoint connection units). Multipoint connection unit servers were introduced for 38 units at seven locations in the Tokyo head office and 106 units at 71 locations nationwide, making a total of 144 units at 78 locations. These enabled a maximum of 80 units to connect simultaneously, and four conferences to be held in parallel. A system enabling conference reservations and hosting from the World Wide Web via the company intranet was also incorporated. Thus the ICT investment, globally prominent at the time, was implemented as a result of the president's top-down decision making.

The staff at the head office's general affairs department drew up an ICT strategy as a confidential business plan, allocated an investment budget for that purpose, and distributed circulars throughout the company. Then the workplace organization staff was addressed to encourage them to apply VIN dynamically in line with three demands from the CEO. A deliberate strategy was thus implemented from a head office overhead subject, as in Chapter 3's pattern one. As a result of implementing the deliberate strategy from head office guidance, however, the VIN had a low operating ratio and failed to become effective. Head office's initial aim of holding around five meetings a day (25 per week) in reality became 0.4 meetings a day, or around two per week, with an operating ratio of around 7%. Thus neither the VIN investment effects nor the fruits of the three demands initially planned by head office were realized. The formulation and implementation of deliberate strategy from a top-down approach creates a shared context among all employees through univocal boundaries communication (meaning communication spanning various organizational and knowledge boundaries within the company) among organization members, and builds a consolidated strategy perspective within the company. When deep-rooted equivocality of interpretation and context exists in the workplace organization, however, it becomes difficult to share common strategy perspectives with all employees from head office guidance.

In Sony's case, a negative reaction arose from an equivocal interpretation on the workplace organization side with regards to strategic content resulting from decision making as deliberate strategy among head office departments. The ICT implementation from deliberate strategy then induced inertia among the workplace organization actors. One of the reasons that the top-down ICT implementation failed was the existence of these equivocal interpretations and contexts among the head office and workplace organization sides.

So what kind of negative thoughts and behavior arose among the workplace organization staff? One was a sense of uncertainty regarding new ICT, giving rise to questions and comments such as "I wonder what an IP-based videoconferencing system is?" "IP looks kind of difficult," "What do we do if trouble occurs during a conference?" or "How can we go about having a conference?" At first, during initial implementation resulting from deliberate strategy, a strong sense of uncertainty remained among a large number of employees, despite the head office division distributing standard operating manuals and holding explanatory meetings.

The head office VIN implementation staff picked up on various issues and problem points that employees found to have penetrated the workplace organization, and took initiatives to promote VIN applications together with the workplace organization employees. Implementation of the deliberate strategy guided by the head office was temporarily halted, and a trial process begun with workplace organization employees to create VIN implementation strategies through a process of trial and error. This corresponds to the process of formulating and implementing emergent strategy. The first actions taken by the VIN implementation staff were to dispel the uncertainty regarding VIN among workplace staff. Frequent videoconferencing study meetings making actual use of VIN were held to do this.

All employees were shown actual videoconference images. It was important to grasp the ease of use and the good image and voice qualities by seeing and experiencing the actual images in a study meeting. Then VIN operating methods were explained in detail and the VIN system's ease of use was promoted. Through this kind of study meeting among the staff at the workplace organization, employees formed a shared opinion from their experience of image and voice quality and ease of use. These processes were steps by which workplace organization members enhanced

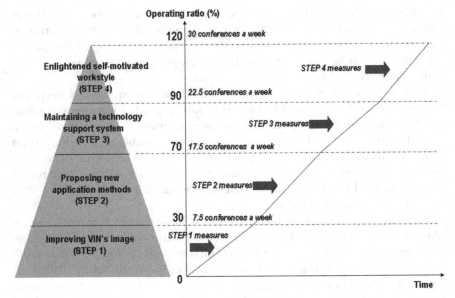

Figure 7.1.   Rise in VIN Operating Ratio Through Emergent Strategy

their "experience value" through "learning by doing." As a result of these processes, the VIN operating ratio jumped to around 30% (see Figure 7.1).

The second step involved proposing a new conference style to workplace staff through study meetings. These went beyond routine conference experiences of morning pep talks, explanatory meetings for introducing new products, supervisor meetings, policy meetings, and general sales manager meetings to include proposing new VIN application methods via morning meetings, explanatory meetings for new systems, ISO environment meetings, safety and hygiene committees, health consulting, and personnel interviews, and included specific implementation focused on high-end (lead) users. These processes led the VIN operation ratio to jump to around 70% (see Figure 7.1).

The third step planned and implemented the building of technology support platforms aimed at further improving the operating ratio. Full measures for a technological backup system were implemented in order to build a "network of trust" aimed at ICT support for employees. Specifically, this involved the creation of a technology backup support center and a VIN application manual to share VIN technology data and enable anyone to use VINs with ease. A call center system was put in place to provide phone support in the unlikely event of trouble occurring during a conference.

With these measures, employees were able to conduct the conference without anxiety. As a result of this process, the VIN operating ratio rose to about 90% (see Figure 7.1).

The fourth step involved specific measures aimed at developing the self-motivated work style of the employees. As one of these measures, the employees maintained the various tools to enable them to build a VIN community independently. Specifically, web-study and an in-house VIN learning system were put in place. As a result, VIN communities were formed through the activities not just of high-end users (lead users) within the company, but of large numbers of employees promoting activities to raise standards for low-end users hesitant to make use of VIN. These processes raised the VIN operating ratio to 120% (see Figure 7.1).

The four steps mentioned above raised the VIN operating ratio dramatically. Then, in addition to the top management's three demands for raising business process efficiency (speeding up decision making, reducing costs, and activating in-house communication), the creation of new VIN-using communities transcending the frameworks of organizational boundaries among divisions and knowledge boundaries among specialist fields was promoted, enabling the inspiration and creation of new knowledge to advance. Moreover, by mastering their own cutting-edge VIN tools, the users' expertise and skills "knowledge" accumulates further, and exploiting this experience strengthens the VIN products' proposal power (such as sales of own-company VIN products) to users outside the company. In this way, VIN contributes not just to business process efficiency but to improving the creativity of knowledge comprising new business.

Thus Sony succeeded in introducing VIN systems by formulating and implementing emergent strategy through trial and error involving the workplace organization, centered on VIN installation staff. Then top management later appended meaning to the fruits of this emergent strategy implementation, and created consolidated ICT strategy perspectives for all employees. The president promoted the strategic importance of VIN tool activities inside and outside the company through press releases, stating that "Our company is acquiring a strategic advantage by introducing an IP videoconferencing system throughout the country ahead of its rivals." These actions of top management embedded VIN activity as business process routines based on consolidated ICT strategy perspectives, and not only pursued business efficiency through VIN aimed at new business innovation but also publicized the major aims of enhancing creativity through such initiatives as using VIN to market new products. Then, regarding future VIN investment, a new start was made to implement strategies systematically and rationally from a second deliberate strategy.

ICT DYNAMIC CAPABILITY
AND NEW KNOWLEDGE CREATION

When arranged in order, the three constituent elements of the ICT dynamic capability in this case—ICT application, context architect, and boundary consolidation—appear as in Figure 7.2. At first, the president's top-down, systematic VIN introduction, considered from the viewpoint of the strategy-making process, was the implementation of deliberate strategy. The top-down creation of consolidated ICT strategy perspectives, however, ended in failure. The feeling of uncertainty from the workplace employees with regard to VIN activity increased the equivocality among employees, with the result that ICT application capability hardly accumulated at all. From this sense of uncertainty, new contexts failed to arise from VIN activities among employees, and almost no division-spanning human networks were created. As a result, new knowledge was not generated among employees.

Learning from the failure of the deliberate strategy led by the head office organization, however, the implementing of emergent strategy centered on the head office staff division and workplace organizations came to enable the formation of new contexts and human networks. Initially, in Step 1, VIN-related study meetings dispelled the feelings of uncertainty

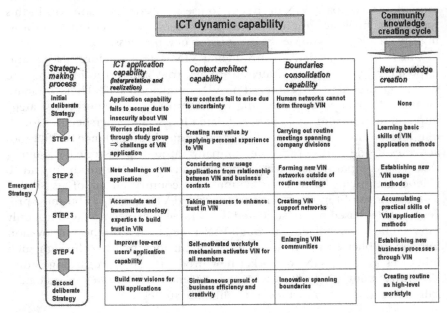

Strategy-making process	ICT application capability (Interpretation and realization)	Context architect capability	Boundaries consolidation capability	New knowledge creation
Initial deliberate Strategy	Application capability fails to accrue due to insecurity about VIN	New contexts fail to arise due to uncertainty	Human networks cannot form through VIN	None
STEP 1	Worries dispelled through study group ⇒ challenge of VIN application	Creating new value by applying personal experience to VIN	Carrying out routine meetings spanning company divisions	Learning basic skills of VIN application methods
STEP 2	New challenge of VIN application	Considering new usage applications from relationship between VIN and business contexts	Forming new VIN networks outside of routine meetings	Establishing new VIN usage methods
STEP 3	Accumulate and transmit technology expertise to build trust in VIN	Taking measures to enhance trust in VIN	Creating VIN support networks	Accumulating practical skills of VIN application methods
STEP 4	Improve low-end users' application capability	Self-motivated workstyle mechanism activates VIN for all members	Enlarging VIN communities	Establishing new business processes through VIN
Second deliberate Strategy	Build new visions for VIN applications	Simultaneous pursuit of business efficiency and creativity	Innovation spanning boundaries	Creating routine as high-level workstyle

Figure 7.2. ICT Dynamic Capability and New Knowledge Creation

among the workplace staff, leading to productive interpretation and creative realization of VIN among the staff and inspiring them to use VIN in new challenges. As a result, the new context of new value created from VIN activity through practical experience at study meetings enhanced the workplace staff's context architect capability. The workplace staff further responded to new contexts by using VIN to carry out routine meetings spanning divisional boundaries, thus enhancing their boundaries consolidation capability. Then the workplace staff enhanced the three capabilities, acquiring ICT dynamic capability and learning the basic skills of VIN application methods as new knowledge.

As shown in Figure 4.2, moreover, in the individual processes from steps 2–4, workplace organization staff simultaneously enhance ICT application, context architect, and boundary consolidation capabilities while acquiring new ICT dynamic capability in a process similar to that of step 1. With the second deliberate strategy, moreover, top management later added meaning to the fruits of emergent strategy (improved VIN operating ratio), and built consolidated ICT strategy perspectives for all employees. Then the creation of a new vision with regard to VIN activity enhanced ICT application capability; the context of the challenge to simultaneously pursue business efficiency and creativity aimed at further innovation enhanced the employees' context architect capability; and innovatory activities spanning divisional boundaries enhanced boundaries consolidation capability. After that, the employees created a work style of routine accelerated by the new knowledge of VIN activities.

THE ICT DYNAMIC CAPABILITY LOOP AND THE COMMUNITY KNOWLEDGE CREATING CYCLE

In this case, actors' VIN activities led to the dynamic formation of new contexts (conversion of existing contexts), and the creation of formal and informal human networks. These dynamically formed contexts and human networks provide new meaning to actors, and the actors go on to create new knowledge aimed at shared strategy objectives. Actors deliberately (or unintentionally) span organizational and knowledge boundaries through VIN activity, and the context architect capabilities of actors inspired by new strategy perspectives produce new contexts among actors. These new contexts then create boundaries consolidation capability aimed at the building of new human networks among actors. VIN activities promote actors' context architect and boundaries consolidation capabilities in addition to further VIN activity.

Then the new strategic perspective of the management leaders with regard to ICT is to encourage the dynamic ICT activity of new interpretations and realizations. Meanwhile, the actors enhance context architect capability to produce new contexts and boundaries consolidation capability to build human networks, while dynamically interacting with strategic and organizational contexts within and outside the company.

Next, the interaction of context architect and boundaries consolidation capabilities further promotes VIN activity and enhances actors' ICT application capability. ICT dynamic capability then arises spirally as these three elements (context architect, boundaries consolidation, and ICT application capabilities) interact (see Figure 7.3). I have called this the "ICT dynamic capability loop." In the case of Sony, an ICT dynamic capability loop is realized chronologically through the processes of implementing deliberate strategy, once again following steps 1–4 as shown in Figure 7.3.

Implementing the ICT dynamic capability loop, meanwhile, produces new knowledge from the community knowledge creating cycle. The interactive relationship between the implementation of the ICT dynamic capability loop and the community knowledge creating cycle (SICA model: sharing, inspiration, creation, accumulation) can be explained as below (see Figure 7.4). ICT application capability promotes dynamic ICT activities. Perceiving this process from the accumulation and sharing stage of the community knowledge creating cycle, actors share productive interpretation and creative realization among themselves. As a result, ICT activity

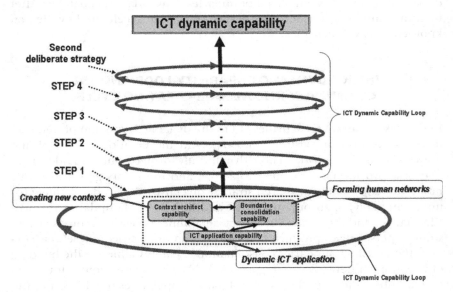

Figure 7.3. ICT Dynamic Capability Loop

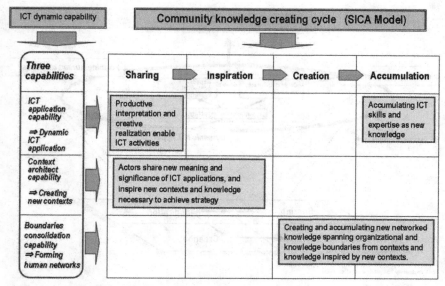

Figure 7.4. The Relationship Between the SICA Model and the Three Elements of ICT Dynamic Capability

reverts to actors' capabilities, based on knowledge that is already accumulated (although in some cases, knowledge necessary for ICT activities has not accumulated within the organization). Next, context architect capability goes to create new contexts. When this process is perceived from the sharing and inspiration stage of the community knowledge creating cycle, the new meaning and significance of ICT activities is shared among actors, and actors revert to inspiring new contexts and knowledge necessary to achieve the strategy. Moreover, boundaries consolidation capability forms new human networks. When this process is perceived from the creation and accumulation stage of the community knowledge creating cycle, context and knowledge inspired by new contexts span organizational and knowledge boundaries, and networked actors return to creating and accumulating new knowledge.

As explained above, implementing the ICT dynamic capability loop links to the community knowledge creating cycle, and promotes new knowledge creation and accumulation in the community (organization). Interpreted in another way, the community knowledge creating cycle process (through ICT activity) in real and virtual space interacts with the ICT dynamic capability loop process to create new organizational capability and competitiveness as an organization applying ICT (see Figure 7.5).

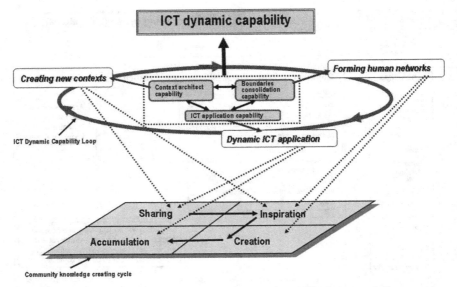

Figure 7.5. Relationship Between ICT Dynamic Capability Loop and Community Knowledge-Creating Cycle

CHAPTER 8

MANAGING PARADOX THROUGH DIALECTICAL MANAGEMENT

As I explained in Chapters 4–7, the process of introducing and applying ICT indicated a great reliance on a strategy-making process that differed from the strategic and organizational contexts within the organization. A uniform pattern of success does not generally exist for promoting the introduction and application of ICT; it always depends greatly on the environment the company is facing and the company's strategic and organizational context. The process of introducing and applying ICT grows more difficult the larger the company becomes (as the organization swells). The reason relates to the different adaptation patterns that individual actors can accept for the ICT applications, alongside the increasing boundaries communication and equivocality of individual interpretations among actors.

The companies and organizations indicated in the case studies in Chapters 4–7, however, reduced equivocality within organizations and enhanced actors' ICT application capability through management of the appropriate strategy-making processes, enabling the acquisition of ICT dynamic capability as a result. This chapter presents two case studies of VIN system introduction in large companies. The knowledge gained from these case studies indicates that the presence of strong links between top and middle management, and the deep dialog and collaboration within and between each management layer, promote dialectical management at the company. In this chapter, I consider the dynamic processes by which actors simulta-

New Knowledge Creation Through ICT Dynamic Capability, pages 151–172
Copyright © 2008 by Information Age Publishing

neously managed the different strategy-making processes of deliberate and emergent strategy; effectively reduced equivocality among organizations and actors; and acquired ICT application capability and organizational ICT dynamic capability as a result of this dialectical management.

BROADBAND IP VIDEOCONFERENCING INITIATIVES: A CASE STUDY FROM OTSUKA CORPORATION

Otsuka Corporation is one of Japan's leading ICT system integrators. Up to now, Otsuka Corporation has achieved much by delivering new business opportunities and management reform measures arising from data and communication technology innovations, in specific forms to a large number of companies, and has offered general support to corporate activities. Otsuka Corporation also has one of the best records in Japan with regard to VIN system activity, with a history of VIN system processes spanning about 20 years.

Step 1 Initiatives

During Step 1 (from 1986), videoconferencing systems were introduced to the head office in Tokyo, the development division in Chiba, and the Osaka branch office, and dedicated videoconferencing rooms were set up, mostly for directors meetings. The company established dedicated high-quality, high-volume, and high-speed digital lines, and invested in ICT to the tune of hundreds of millions of yen. The cutting-edge devices of high-speed videoconference systems were introduced ahead of rivals. Top management put into practice its own activities, and held an appeal for corporate customers as a company with advanced ICT application. It was also important that the dynamic activity of top management encouraged communication with in-house middle management and lower levels. Top management's intention in a large organization was to present a model within and outside the company as an ICT solution company, and at the same time speed up decision making and activate communication spanning organizational boundaries within the company as goals of a major organizational strategy. In Step 1, the consolidated ICT strategy perspective was built and shared within the top management team. Then, top management, starting with itself, introduced VIN tools in-house as a result of rational and analytical deliberate strategy, and implemented VIN tool proposals as ICT solutions for corporate customers.

Step 2 Initiatives

Around 7 years after Step 1, in Step 2 (from 1993), top management launched full-scale innovations in organizational culture within the company. The videoconferencing system installed previously in Step 1 had adopted a model equipped with what was then the latest technology, and Otsuka had become one of Japan's few large-scale users, with more than 60 videoconference system units.

Then Otsuka banned business trips for the purpose of attending conferences for all employees, and promoted in-house videoconferencing activities. At that time, a somewhat negative reaction developed among middle management, focused on the workplace, regarding top management's intentions. However, the ICT project's middle management team from the head office division, whose mission was to advance close dialog and collaboration with staff at the workplace organization with regard to VIN activities, communicated top management's univocal interpretation to the workplace organization effectively and promoted VIN activities throughout the company. Then dialog and collaboration between the workplace organization and the head office centered on middle management, reduced the equivocality of the actors' individual interpretations of VIN introduction and activities, and actors built and shared consolidated interpretation among head office and workplace organizations. These dialog and collaboration processes were also implemented through trial-and-error actions relating to numerous methods of use and VIN environment maintenance. These activities did not simply involve one-way equipment delivery from the head office side, but incorporated plans for new activity methods and customizing tools suited to the workplace environment.

Through this process of trial and error, the structure for a range of virtual conferences was created within the company. A large number of VIN use formats were created, not just for formal conferences but also for training and informal study meetings, and for opinion exchanges to share information. Thus a large number of new usage formats were proposed that the top management of the time had not predicted. These came about partly from the implementation of emergent strategy focused on the middle management layers at the head office and workplace organizations. But the management ranks, especially the CEO, gradually propagated VIN usage opportunities within the company, and created an atmosphere prevailed where it was felt that work could not be properly accomplished in the future without VIN tools, thus enlightening others about VIN tools. If the only action had been to ban business trips, the employees' adverse reaction alone might have ended the initiative, but in Otsuka Corporation's case, because it accompanied the shifting of the corporate landscape toward VIN activity, the sense of conferences as business trips' "just cause" weakened naturally, and two birds were killed with one stone as decision making accelerated and transport costs fell.

Step 3 Initiatives

Then in Step 3 (2003 onward), Otsuka Corporation embarked on installing IP through broadband in all conference networks within the company to coincide with the completion of the new head office. Steps 1 and 2 had relied on ISDN as the telecommunications line, and as the conference hosting time and connection bases rose, Otsuka was paying around 20 million yen a month to a telecommunications company for connection charges. Top and middle management had to somehow find a way to improve this figure. Then the project to convert the videoconferencing system to IP became a middle management project centered on the head office. One of the missions assigned to the project was to slash ISDN operational costs to 8 million yen a month by converting to IP. At the same time, it became important to demonstrate an accurate grasp of operational costs by reducing expenses. The second issue was to realize a VIN system possessing clear, realistic images and voice. Devising a means of establishing more realistic video suited to executive conferences, with TV-quality resolution and sound, alongside long-term, stable operations, became a major topic. The third issue was improving operability while reducing the administrative burden of the operation. Major issues to accomplish this involved introducing a user-friendly videoconference system to increase frequency of use, converting to easily manageable terminals, and reducing the administrative burden of the head office's general affairs department.

The project team made progress investigating these technological, maintenance, and financial issues through trial-and-error. On the technology side, the thinking concerning the communication zone of the Internet VPN (virtual private network) environment was reconsidered as a major issue. Specifically, the problem involved risks associated with the remarkable fall in sound and image quality in the Internet VPN operating virtually in real time, including videoconferencing systems and IP phones, and the coexistence with the normal intranet environment. As a countermeasure, the general data and voice-video networks were physically separated. Another important issue was devising a means of stable operation. The Internet environment described above cannot guarantee the reception of stable service, requiring dual pathways for critical operations and a backup system. It is also necessary to fully investigate issues such as mechanisms and support systems to counter attacks through security weaknesses. As countermeasures, the project team investigated network backup circuits, and determined that monitoring of firewalls, such as duplicating network devices, and remote maintenance service activity was effective.

Then, as a result of investigating detailed videoconferencing network infrastructure, the decision was taken to merge voice and data communications for core networks within the company. One reason was there were no examples anywhere in the world (and there are none today) of adding videoconferencing to core networks within a company focused on data communications. Another was the decision that if videoconference networks were to merge sound and data communications, it was appropriate to consider separating them from the core network, which has a large number of unstable elements with regard to voice quality. The decision was taken to build an independent, closed area network as a VIN system. At that time, the emergence of ADSL as a videoconferencing system access circuit began to enable the cheap use of comparatively high-speed lines. The ADSL service area was narrow, however, and there was no guarantee of voice quality or stable regional provision. For that reason, the then cutting-edge technology of optical fiber was adopted for the Internet connection.

By solving the kind of problems mentioned above, it then became possible to expand the installation activity across divisional units by completely switching IP models through broadband, and further, to expand connection bases to consolidated companies and to the business office in Shanghai, China. Then in Step 3, by expanding the scale of VIN installation to 66 units in 30 bases throughout Japan and increasing multipoint connection units to 48 bases with two units each, installation costs were kept down to 2 million yen per unit and running costs to 3 million yen a month, and a communication network running completely under IP was realized (see Figure 8.1).

As a result of introducing the VIN system through trial and error in Step 3, the VIN activity effect was acquired as described below. One aspect was the simultaneous broadcast (effect of reducing assembly costs; the morale-raising effect communicated by the sense of presence and expectation that differs from reading out instructional documents) as an all-company event (including ceremonies to commemorate the founding of the company and New Year greetings from the CEO). Another was the participation of 200–300 executives in monthly events including regular reports and strategy conferences. This reduced costs for all executive business trips to regional bases and transportation expenses for executives within the Tokyo region, and it also enabled overseas workers to get a timely sense of how head office operates. A third aspect was the participation of between 50 and 100 people in each division responsible for holding divisional meetings 20 to 40 times a month. This not only cuts down on transportation costs but avoids loss from the absence of departmental chiefs and enables the smooth communication of new information. A fourth aspect was the

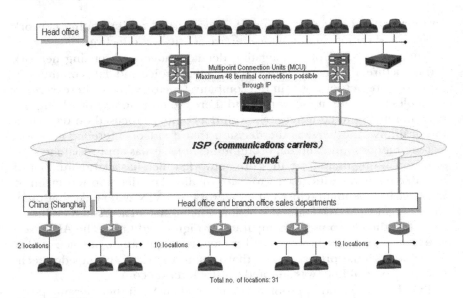

Source: Compiled from interviews.

Figure 8.1. Osaka Corporation's VIN (Step 3)

acquisition of major benefits: research and study meetings can be held at any time, a useful means of smooth, fast data communication has been gained, and there is no need for a large venue to assemble study groups of salespeople and engineers. Fifth is the enhanced use for project discussions and discussions among people responsible for different divisions at any time. The use of images in real time also increases the level of understanding for content that is difficult to communicate by phone.

Step 3 enabled a rise in VIN activity extending to workplace organization staff members lower in rank and ICT maturity than those involved in Step 2. In Step 3, as in Step 2, trial-and-error action took place through the implementation of emergent strategy, centered on middle management, spanning the head office and workplace organizations. Based on univocal, consolidated ICT strategy perspectives from top management in Steps 1–3, the equivocality of boundaries communication and interpretation at the middle management and lower levels gradually decreased through dialog and collaboration, and consolidated ICT strategy perspectives were shared among all employees through a process taking around 20 years to move from Step 1 to Step 3. Then the ICT application capability increased among all employees and led to the creation of ICT dynamic capability over the whole of Otsuka Corporation's corporate organization.

INITIATIVES TO CHANGE CORPORATE CULTURE: A CASE STUDY FROM NTT

Changing from a Telephone Company to a Multimedia Company

On New Year's Day 1994, then-president Kojima declared that NTT would change from a telephone company to a multimedia company. At the time, NTT was facing a significant transition period from an analog telephone business with a 40-year history to a future of creating new business using multimedia.

Following privatization in 1985, NTT implemented incremental change by spinning off group companies (the data communications section in 1988 and the mobile communications section in 1992), entering new business fields, introducing voluntary retirement, and carrying out other in-house streamlining policies. As a result, by the beginning of 1994 the company had reduced its personnel numbers from around 300,000 to around 180,000. Income from analog telephone sales, the core of the company's business, was gradually falling due to the entrance into the market of new common carriers and reductions in telecommunication fees accompanying liberalization of the telecommunications market in 1985.

On the other hand, the demand for nonvoice services such as data communications was gradually increasing, primarily from corporate users. At that time, however, this was an unknown field for telecommunications carriers in terms of what strategies should be used to create what kind of services for the Internet, which had begun spreading, primarily in the United States, from 1994.

Construction of New Organization by Top Management Members

The two persons who felt the greatest sense of crisis regarding NTT's future were then-president Mr. Kojima, who made the "Multimedia Declaration" in 1994, and then–vice president and person responsible for technology Mr. Miyazu (who took up the president's post in 1996). They were searching for a future structure for the business through radically changing NTT's constitution (including its organization culture and business style). They reached the conclusion that they had to construct a new organization in the head office to introduce multimedia strategies for the future. The person who was assigned this responsibility was then–board member and director Mr. Ikeda (he became managing director in 1996). This organization was named the Multimedia Business Department (MBD) and was started in June 1994 with about 50 staff members. Two years later, the department had increased to about 850 staff. Top management members Mr. Kojima, Mr. Miyazu, and Mr. Ikeda shared a sense of values and future vision and had a firm belief and will to dismantle NTT's traditional organization culture and create a new multimedia business market that drew from information and communication technologies.

With respect to the multimedia organization to be created, Mr. Miyazu instructed Mr. Ikeda to create a structure where people could do new things based on new ideas. Discussions among top management led to a shared sense of values and vision of developing from a telephone company to a multimedia company, enabling Mr. Ikeda to begin creating a new organization, which became known as MBD. Making good use of his experience as a former general manager of NTT's personnel division, Mr. Ikeda gathered various staff from among the remaining 180,000 employees, picking up, for example, researchers and engineers who were each credited with more than 100 patents or utility model rights, young staff with entrepreneur experience during their school days, and top system engineers in the Kansai area. The project leaders who were to lead the new MBD organization were selected from an upper-ranking branch manager class who had a good sense of balance. At peak, more than 850 staff, from veterans to capable young personnel, were formed into teams.[1]

The overriding characteristics of the new organization were the degree of freedom, speed, and tension all members experienced, which was embodied in the shared feeling that "of the entire NTT organization, only our group (MBD) is not NTT." MBD differs greatly from NTT's traditional existing line organization, which has about 170,000 staff. The income from analog telephone services was obviously NTT's core business at that time, and line employees who engaged in this business formed and retained the organization culture, including the existing business processes and customs, and planned and executed extremely detailed and systematic deliberate strategies related to telephone service (strategies were planned in each division by function, such as sales planning, facility planning, customer service, and maintenance). They have continued providing reliable services to customers throughout the nation for the past 10 years. Incremental reform took place in day-to-day activities, and in this disciplined line organization, body information and knowledge was handed down and shared in a top-down method, from the head office to branch offices and from branch offices to sales offices (hereinafter, the existing line organization is referred to as the traditional organization). When MBD was promoting unique businesses based on new ideas, severe conflict with the traditional organization was unavoidable.

At the moment a new organization or system is created, it is always already behind the times. To start a new thing, the existing systems and ideas must be changed and people must be energized. As time passes in an environment characterized by habitat segregation of changing industries (such as fusing communication and broadcasting), NTT must change as well. It is only natural that there will be intense arguments at that time.[2]

Because of this, the sharing of data and knowledge became the most important issue among MBD and the traditional organization. Then it became necessary to move forward by applying information sharing to ICT tools through real-time communication with workplace organizations, which are traditional organizations dispersed between the Tokyo MBD and offices throughout Japan. Constructive and productive conflict was induced between MBD and the traditional NTT organization through top-down transmission of the vision of top management and the corporate-wide promotion of knowledge management utilizing ICT tools between mutually conflicting groups.

Development of Yarima SHOW Multistrategy

The composition of the frontline sales branch offices rapidly moved away from the analog telephone-type business framework under the direction of the head office and the execution of routine work processes. In 1996, using the "Yarima SHOW Multistrategy" management innovation project, which espouses the principle that once change has begun, there is no going back to how things were before, an effort was launched to convert all 200 branch offices nationwide to multimedia branches. This change had two strategic aims. One was a change in business style, wherein business at branch offices switched from phone to multimedia. The other was a change in work style, where the work methodology itself is transformed through the use of multimedia tools, such as intranet, videoconferencing, and groupware (see Figure 8.2).

At the same time, this activity was positioned as a "revolutionary movement," for two reasons. One was that the existing telephone and multimedia businesses are totally different worlds in terms of product and service characteristics and the sense of business values. Here, the word "revolution" means that the conventional values were discarded to create a totally different new business field for NTT. The other reason related to the quality of the competition NTT faced. At that time, many companies were about to enter the market, including computer manufacturers, electric appliance manufacturers, communication companies, and electric power companies. This influx was to rapidly turn the multimedia market into a competitive arena that differed completely from the environment NTT had operated in to date, and results had to be obtained quickly. This revolution was a "revolution of speed" that required that the gradual method of

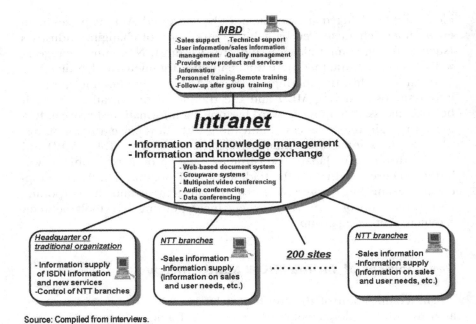

Source: Compiled from interviews.

Figure 8.2.   ICT Tools for Knowledge Management through Yarima SHOW Multi Strategy

change used in conventional reform be switched to one of dramatic change. This totally different approach was required, since the essence of the requested revolution was fundamentally different from what had happened before.

Three Restrictions Lifted by ICT Tools

Time: Faster progress to execution stage. The most important characteristic of ICT tools is that they can be configured quickly and at low cost. In the case of the "Yarima SHOW Multistrategy," the first website was set up just a month after the project was started and information dispatch to all branches began. This speed is beyond the imagining of the conventional system of network configuration. Using these tools, the revolution was accelerated.

Conventionally, when NTT begins measures involving the entire company, notification is first sent from the head office to branch offices and adjustments are made between the head office and branch offices, and between branch offices. Then adjustments and announcements are made inside each branch office. With this approach it took a long time to actually execute measures. On the contrary, with the "Yarima SHOW Multistrategy," every possible effort was made to eliminate the conventional pyramid-type information route whereby information is passed from the head office to branch offices to sales offices, instead choosing to exchange information directly between the head office and branch or sales offices. Additionally, with emergent processes such as "think on-the-fly," the basic stance was that decisions made at the head office were dispatched in real time via ICT tools, and branch and sales offices were expected to react. This created a revolutionary style of executing a decision immediately, using multimedia tools to dramatically decrease the lead time required from decision making to execution process.

Distance. It was very difficult for the 850 staff of MBD alone to move all branch and sales offices in an extremely large company such as NTT. When there was almost no reaction from branch offices at the beginning of the project, MBD analysis indicated that this was "evidence that the revolution has not started yet," which was a natural outgrowth of MBD's belief that "results cannot be obtained if the action is not recognized. There are always opposing opinions during a revolution." When the intranet website was launched and information dispatch became active, the number of inquiries and requests from branch offices and sales offices began to rise. Responses to outstanding activities introduced on the Web page began to come in, and some branch offices volunteered to dispatch information on their own initiative. This indicated that the intranet was functioning as an engine of revolution to accelerate the recognition from activity to dissemination, and from dissemination to a series of expansive actions, and thus became a trigger of the revolution. President Miyazu instructed the company that "multimedia-related information must be conveyed from the head office to the person responsible in each branch equally in terms of the amount, quality, and level, regardless of their position." ICT tools were used for this purpose as well.

Form: Flexible revolution beyond the existing framework. Because of their flexibility and expandability, multimedia tools really demonstrate their effectiveness when the speed of change is rapid and it is difficult to foresee the future. In such cases a flexible response is necessary for a revolution that is changing on a minute-by-minute basis. ICT tools are also suitable for promoting a revolution free from the conventional framework. Past reform movements within NTT were led by the head office and the traditional organization, and the success of the revolution was based on group responsibility guidelines. There was also a strong sense that it was necessary to first create an organization and establish a system, so that organizational reform took precedence. Decisions were therefore institutionalized and sometimes grew obstructive when it became necessary to correct the direction of the revolution. The means of communication was mostly paper-based, including notification documents and business manuals, and importance was placed on a methodology whereby people met physically in such settings as nationwide caravans or meetings.

The main thrust of the "Yarima SHOW Multistrategy" was always branch offices and the self reform of branch offices through their own initiative. The greatest priority was placed on taking action, and for that purpose they did not mind overriding existing rules. The stance was that the framework of the organization was later changed based on achievement. For that reason, to encourage reform in the actions of responsible personnel in each branch, multimedia tools were utilized as much as possible as revolutionary tools that substituted for the conventional paper based decision-making process or face-to-face meetings. Information was dispatched directly and in real time. For example, when one branch came up with a successful sales method, other branch offices were asked to do the same thing.

ICT tools such as intranets and videoconferencing systems were used as much as possible as a lever for a revolution prioritizing speed and achievements, based on the idea that one "should change actions right away and get results," unlike the conventional way of thinking that espoused the idea that one should "change consciousness first, then change one's actions." Employees could also become familiar with ICT tools through personal, hands-on exposure.

Since all employees tackled the transition to a multimedia company together, the traditional organization began to recognize that they could obtain great advantages such as expanding business areas, improving awareness among employees, and cultivating talented personnel. In this way, dialogue and collaboration between both organizations were increasingly promoted. A series of activities promoted by MBD revolutionized the consciousness of frontline employees throughout Japan in the traditional organization. They became actively engaged in multimedia, thus disseminating the principles of new multimedia services.

Meeting of 100,000 People

Meanwhile, a "100,000-person meeting" of NTT organizations and employees was held using multimedia tools (Nikkei Sangyo Shimbun, 1998). This meeting was more than just a launch ceremony. The real intentions were to:

- Have employees feel a part of the multimedia movement and recognize that the organization is undergoing conversion from a telephone to a multimedia company.

- Recognize the real power of the NTT in-house network through simultaneous meetings of units of 10,000 employees.
- Confirm NTT's multimedia-related technologies and accumulate expertise.

The "100,000-person meeting" was a new era in terms of the evolving consciousness (motivation or elevation of motivation) of all employees and the cultivation of talented personnel through acquiring skills via actual hands-on experience with multimedia technologies.

INTEGRATING DIFFERENT STRATEGY PERSPECTIVES

The shared knowledge acquired from the two case studies develops as below. One aspect is the coexistence of different strategy perspectives within the company and the fusion and integration of these perspectives through changing times. The consolidated, univocal ICT strategy perspective formed by the top management team (at Otsuka Corporation, this comprises the management ranks led by the CEO; at NTT it comprises CEO and president Mr. Kojima and his successor, Mr. Miyazu; and at MBD it comprises the managing director, Mr. Ikeda) becomes a common strategy after the formation of shared values among the middle management team of the head office organization (at Otsuka Corporation this comprises the project team promoting VIN installation, and at NTT it comprises the project teams within MBD), and the deliberate strategy formulated by the head office organizations is implemented (see Figures 8.3 and 8.4).

A second aspect is the existence of mixed task teams of middle managers from head office and middle managers and staff from the workplace organizations. A variety of equivocal dialog and interpretation took place among individual actors within the task teams with regard to univocal ICT strategy perspectives consolidated by the head office. These highly equivocal interpretations generated diverse opposition and conflict within the task team. An important aspect of the mission of the head office middle managers regarding this task team was to properly communicate the meaning and significance of consolidated perspectives for ICT strategy throughout the company through dialog, and debate constructively with the actors. In these case studies, dialog aimed at building and sharing consolidated strategy perspectives continued endlessly at the workplace. By implementing the emergent strategy of trial and error, actors discovered new value in VIN tools from a process of learning through experience. As for employees with negative attitudes, middle managers at the head office dynamically provided a venue giving the opportunity for experience.

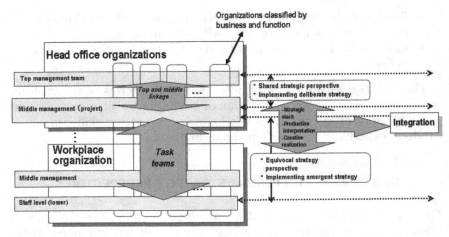

Figure 8.3. Leaders' Communities and Strategy Structures for Each Management Layer: Case Study from Otsuka Corporation

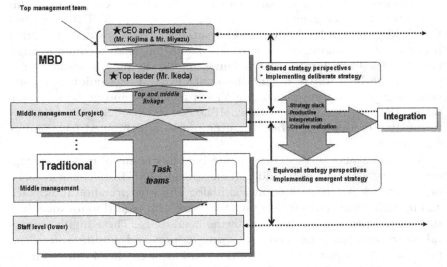

Figure 8.4. Leaders' Communities and Strategy Structures for each Management Layer: Case Study from NTT

The integration of different strategic perspectives involves implementing deliberate strategy from consolidated, rigid strategic perspectives while implementing or permitting strategies from various perspectives that strengthen (or complement) the original strategies. The actors' emergent action arising from the permitted variety of strategic perspectives becomes implemented as emergent strategy in the workplace organization. The integration of deliberate and emergent strategy can be interpreted as follows. Amid the implementation process of ICT strategies rigorously formulated by the head office organization, the response to environmental changes in the workplace organization (such as customer support, changes in the service level, and local improvised action) corresponds to top management's consent for "strategic slack" (the elastic degree of strategy freedom). This "strategic slack" produces creative originality of ICT activity as the workplace organization's emergent strategy, and also produces new application structures that the head office had not imagined. Put another way, the workplace organization actors understand the strategic perspectives as "productive interpretation," and while complying with these, produce new ICT activity styles from their own, original methodologies from creative realization catalyzed by strategic slack. To integrate this kind of strategy perspective, top management and management leaders must permit strategic slack among the employees, and enable productive interpretation and creative realization. The resonance of value process among actors is important in bringing this about, as shown in the next section.

BUILDING A PLATFORM OF RESONATING VALUES VIA DIALECTICAL MANAGEMENT

In these case studies, it was necessary to promote creativity and resonance of new values among each management layer in order for ICT strategies to be understood by and penetrate to all employees, including the workplace organization, as shown in Figures 8.3 and 8.4. Leaders focused on each management layer (management executives in the case of top management; project leaders and workplace organization middle managers in the case of head office middle managers) had to properly explain ICT application strategy perspectives to other members, based on the ideology and philosophy of the interactive learning-based community (Kodama, 2004), and the company had to accurately convey the ideal purposes and goals of ICT activity. In order to do this, it was important to create new values and resonate these with all organization members.

Each individual employee in each management layer had to strengthen the feeling of solidarity within and among each management layer with "task teams" aimed at creating and resonating the new values of producing new business markets to activate the organization through ICT (Otsuka Corporation) and reforming corporate culture (NTT). Moreover, the leaders in each management layer had to strengthen the sense of solidarity within the corporate community of each organization member (including lower layers) of the head office and workplace organizations (nonformal organizations formed from "task teams" within and among each management layer will be termed "communities" below in this section).

In these case studies, creating new values in the community and forming a platform to resonate new values, including all community members, in order to activate organizations and change corporate culture were important issues. Actors within the community (termed "community members" below) unified their ideas, intentions, and spirit, aiming to realize new values targeting the creation of new business markets as a result of activating the organization (in Otsuka's case) or changing corporate culture (in NTT's case) as declared by top members led by the CEO. Accomplishing these aims did not involve the denial of all Otsuka's and NTT's established values (or their perception as "old values"). Rather, they involved the synthesis of new values with traditional organizational values cultivated by the accumulation of path-dependent knowledge in the past. In the case studies in this chapter, the process of resonating values (sharing -> inspiring -> creating -> resonating) was observed as follows (see Figure 8.5).

The sharing of values, the first step, is a stage in which all members of the community aim to learn and understand the new values that indicate

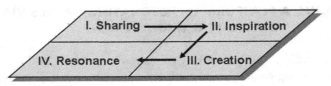

I.   Sharing
      Studying and understanding a new values advocated by top management
      and community leaders, and studying and understanding contradictory
      values and events

II.   Inspiration
      Challenging old values and inspiring a synthesis with new values

      - Problems and issues are confronted, concrete measures are looked into,
      and dialectical thinking and action that resolve conflicts for the sake of
      mutual benefits are inspired.

III. Creation
      Creating new values in communities

      - Contradictory values and events are synthesized.

IV.   Resonance
      Resonating the ideals, will, and spirit of all community members for new
      value in the communities

Source: Kodama(2004)

Figure 8.5.   Resonance Process of Value in the Community through Dialectical Management—Unification and Conflict of Opposites

the direction in which the community believes it must proceed. The community members study the differences between existing values and new values through constructive dialog with top members of management and community leaders (such as project leaders) to understand the essence of the new values. In NTT's case, for example, for community members (especially employees of traditional organizations) that have come to follow a profession in an established phone business, mastering new ICT tools and facing the challenge of new multimedia businesses were great hurdles. However, community members dynamically learned new values and deepened understanding as a result of constructive dialog repeated enthusiastically by top management members and community leaders.

The inspiration of values, the second step, involves challenging old values and triggering ideas within the community aimed at synthesizing the old values with new values. This process involves confronting problem areas and issues within the community and triggering the dialectical thinking of community leaders and members as they search for concrete measures. This step also involves identifying common benefits for all community leaders and triggering the resolution of conflicts. One of the important triggers in this process of inspiration is to arouse a sense of crisis among all community members. In NTT's case, top members of management and community leaders persuaded others of the need to pursue new business structures in order to survive in the coming multimedia age. The business actions emphasized by community members were the promotion of prompt information sharing throughout the company, and the actions of the kind of employees who transmit new information in real time and stimulate new experience.

It is important that this sense of crisis imbues community members with the recognition of new problems and that the innovative leadership of members of top management and other community leaders at each management level synthesizes the old and new values. In NTT's case, the conventional concept of implementing existing business processes and the new concept of a new business style applying ICT aimed at new multimedia business were simultaneously incorporated, and this corresponded to sparking a movement toward synthesis.

In the third step, the creation of values, a synthesis between contradictory values or events is achieved within the community and new values are created. Individual community members aiming for the new objectives of new multimedia business also produce new thoughts and actions within the community by applying cutting-edge ICT and facing challenges through effort. The challenging objective of acquiring new business and technology skills applying ICT within the organization is also established at this stage. New markets are established through new multimedia businesses, and new values are created involving the will and ambition to deliver new value to the customer.

In the final step, the resonation of newly created values, all members of the community resonate aspects of the mind such as philosophies, intentions, and the spirit within the community, and form a sense of unity as a community. Through this process, they form a value resonation platform in the community. Individual community members work at acquiring new knowledge, skills, and know-how while raising their own existing knowledge and competence and recognizing unwavering courage and confidence with respect to the new business utilizing ICT tools. Then the resonating of the values of all employees within the company is promoted, with the aim of driving forward new business.

The process of resonating the values above is especially important to the top members of management and community leaders. This process promoted collaboration and cooperation within and among each management layer (task teams), and converted the conflict within and among each management layer to constructive conflict. As a result, a consensus formed, and it became possible to achieve the merging of the paradoxical strategy formations of emergent and deliberate strategy based on the two-way differentiated organizational culture and core competences of head office and workplace organizations. By persuading others to work together to discover the kind of methods that would bring greater benefits to themselves and to partner community leaders (middle managers of workplace organizations) through the dynamic application of ICT, community leaders (head office project leaders and middle managers) undertook integrative thought and action to deal with emerging conflicts in a forward-looking manner (Schmidt, 1974). Then the organizational management leaders (including top management and project leaders), in a pattern of thought and action helping each other to achieve purpose and goals, appropriately managed conflict within and among each management layer (task teams). Moreover, concerning the elimination of the relationship of conflict among top members of management, including the CEO, and community leaders, dynamic action was taken to search for means to confront the issues and problem areas of both sides (Burke, 1970).

The major aspects that have become clear through this case study are those of dialectical technique (termed "dialectical management" in this book), whereby positive and constructive dialog and collaboration within and among each management layer (task teams) synthesized the paradoxical thought and action patterns among all the community leaders, and simultaneously created satisfaction and integrated the reform of individual actors' work styles and the benefits of ICT activity to the whole company.

Dialectic first appeared in the question-and-answer technique of Socrates and Plato's theory of ideas, and it became an approach to thinking about things that was discussed and developed through the history of philosophy. In particular, Hegel (1967) considered dialectic to be a law of dynamic development in cognition and existence, proposing the thesis, antithesis, and synthesis scheme of logic and the concept of "aufheben" (to sublate). According to Marx (1930, 1967), Engels (1952), and others, Hegel's ideological dialectic developed in a practical methodology. He applied dialectic approaches to thinking to civilization and culture, produced thesis and antithesis with respect to propositions and historical fact, and proposed a methodology by which problem areas and conflicts were resolved through the synthesis of the two sides. The synthesis then became a new thesis, both of which were denied by antithesis, which produced another new thesis in a never-ending process; the process of historical development was proposed to be an eternal process. Dialectic, on the other hand, was also applied to organization theory, stimulating discussion based on absolute truths or morality in devotion to the community (Benson, 1977) or in the process of corporate reform (Van de Ven & Poole, 1995). In addition, Peng and Nisbett (1999) and Peng and Akutsu (2000) analyzed the psychological reactions that could easily result from two apparently contradictory propositions and, while risking crises that allow contradictions, proposed "dialectical thinking in a broad sense" that judged parts of both propositions to be correct. Recently, Seo and Creed (2002) used a dialectical perspective to provide a unique framework for understanding institutional change that more fully captures its totalistic, historical, and dynamic nature, as well as fundamentally resolves a theoretical dilemma of institutional theory.

One of the basic laws of dialectic is "the unification and conflict of opposites."[3] In this concept, opposing tendencies, qualities, and features in the workings of nature, society, or human thinking are not only unified and mutually dependent, but are also mutually exclusive and in conflict. In the midst of this diversity of things (such as problem areas or issues), dialectic is also a methodology that is deeply cognizant of the contradiction of opposites, such as action and reaction or assimilation and dissimilation, and discovers a unity that further transforms this contradiction into a driving force for innovation. In this way, dialectical thinking becomes the driving force that gives rise to movement, change, and development toward innovation as a solution to paradox and conflict. The value resonation process is also a concept that embraces "the unification and conflict of opposites" philosophy. As a result, based on the new values that were resonated with employees, knowledge of high quality is created in the community through a series of processes involving the "sharing, inspiration, creation, and accumulation" of the community knowledge creating cycle described in Chapter 2.

ENHANCING ICT DYNAMIC CAPABILITY
THROUGH SPIRAL RESONANCE OF VALUE

The resonance of value process mentioned above reduces the degree of equivocality among actors and builds consolidated perspective among actors. Strategic perspectives rooted in newly created and resonated values induce productive interpretation in actors with regard to strategy throughout the company. In these case studies, new strategic objectives of workstyle reform activating VIN tools emerged. This resonance of value process does not end after a single revolution.

Amid chronological change, actors execute actions according to the individual situations they face while interacting with strategic and organizational contexts within the company amid chronological and dynamic environmental change. Amid the process of executing these daily routines, actors face various problems and issues, and create new values by applying further creative originality of ICT to overcome these problems. Moreover, actors themselves discover new value through ICT activity based on new thoughts and ideas with "strategic slack" as a catalyst. Then these new values spread among actors via empathy and resonance.

For example, a new VIN application method proposed by certain actors is grasped and resonated among actors, and the process of justifying this new method within the company is also a process of resonance of value. Individual actors discover new issues and problems while dynamically interacting work-related contexts and whole-organization contexts through daily ICT activities. Next, actors discover methods that should solve these problems from practical learning processes. These learning processes are induced by individual actors' "creative realization." Moreover, strategic slack from moderate top management becomes a trigger for producing creative realization from new concepts among middle managers and workplace community members. This creative realization produces distinctive, individual strategy perspectives from individual actors aimed at methods of using still newer ICT tools, and means to master the tools. This creative realization is induced, naturally, because it is based on the existence of an ICT-inspired "productive interpretation."

The new ICT application value from independent strategy perspectives is shared, inspired, created, and resonated among actors through the process of resonance of value. This process naturally creates friction and conflict among actors. Top management leaders, however, integrate differing opinions and ways of thinking while understanding the mutual differences, and undertake organizational behavior to produce synergy effects. Next, they synthesize the paradoxes dialectically with the spirit of the interactive learning–based community, mentioned above.

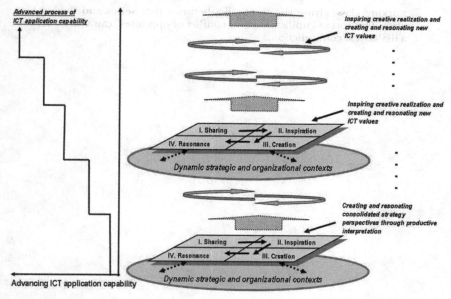

Figure 8.6.   Structure of Platform for Resonating Values Through Dialectical Management and ICT Application Capability

In this way, new, practical behavior as a result of actors' "creative realization" is justified within the group, within the team, and throughout the organization through the process of "resonance of value," and produces new "ICT application capability." Then actors dynamically enhance ICT application capability through the spiral processes of this resonance of value (see Figure 8.6).

NOTES

1.   Refer to AERA (1999), "Gendai no Shozo."
2.   Refer to interview with Mr. Ikeda, AERA (1999), "Gendai no Shozo."
3.   When the relationship between the working class and the capitalist class is considered from the viewpoint of "the unification and conflict of opposites," a capitalist class can certainly exist simply because there is a working class that they hire and exploit. The working class, which does not have a means of production (machinery, plants, etc.), sells their own labor to the capitalist class and has the capitalist class earn a profit (i.e., allow themselves to become exploited) so that they can support their lives. While this alone would make the relationship between these two sides a "unification of opposites," they are at the same time in a relationship of conflict in which their mutual interests can be in opposition. If the capitalist class strives to increase profits, that means lower wages and longer working hours for the

working class, giving rise to conflict between themselves and the working class. This sort of "unification and conflict of opposites" can also be termed a dialectical contradiction.

CHAPTER 9

NEW KNOWLEDGE CREATION THROUGH ICT DYNAMIC CAPABILITY

In this section, I use the crucial aggregated dimensions derived from grounded theory to consider what impact the organizations (including customers) among the knowledge communities (corresponding to the communities of practice, strategic communities, and those networked groups mentioned in Chapter 2) and the ICT dynamic capability arising from the innovative leadership of management leaders on new knowledge creation. First I analyze the characteristics of knowledge communities that triggered new knowledge creation among organizations. These characteristics, which were derived from grounded theory, consist of four broader dimensions, which the author refers to as deep collaboration, dialectical dialog, resonance of values, and building trust. Second, using two broader dimensions that the author refers to as the ICT dynamic capability loop and the community knowledge creating cycle described in Chapters 1 and 2, the author discusses new knowledge creation and ICT dynamic capability through innovative leadership by which management leaders dialectically synthesize the various strategy-making processes (patterns 1–5 described in Chapter 3) within organizations, and then exploit them through the ICT dynamic capability loop and community knowledge creating cycle within knowledge communities.

DEEP COLLABORATION, DIALECTICAL DIALOGUE, RESONANCE OF VALUE, AND TRUST BUILDING WITHIN KNOWLEDGE COMMUNITIES

As mentioned in Chapter 2, actors' "productive interpretation" of ICT is required in order to enhance ICT dynamic capability. Productive interpretation through deep understanding of ICT spanning top and middle management layers promotes actors' ICT activity, and becomes a source of deep collaboration among actors.

Collaboration has been studied in a wide variety of literatures. According to Gray (1989, p. 5), who is often credited with formally launching collaboration theory, interorganizational collaboration is defined as "a process through which parties who see different aspects of a problem can constructively explore their differences and search for solutions that go beyond their own limited vision of what is possible." Learning and innovation literature (including Anand & Khanna, 2000; Larsson, Bengtsson, & Henriksson, & Sparks, 1998; Kale, Singh, & Perlmutter, 2000) emphasizes that collaboration does not just transfer existing knowledge but can also facilitate the creation of new knowledge (see, e.g., Gulati, 1999; Powell, Koput, & Smith-Doerr, 1996). Collaboration might also operate as a strategic tool for gaining efficiency and flexibility in a rapidly changing environment by, for example, targeting customers' fields and needs (Westley & Vredenburg, 1991). Thus collaboration theory, which builds on empirical and theoretical perspectives from a variety of research streams including resource dependence, learning, and strategic management (Gray & Wood, 1991; Wood & Gray, 1991), is useful at interpersonal, intergroup, and interorganizational levels of analysis within knowledge communities, such as the promotion of strategic ICT applications as described in these case studies in Chapters 4–8.

As mentioned in Chapters 6 and 7, actors in head office and workplace organizations continually create methodologies aimed at developing further new uses for ICT arising from collaboration through deep dialog and trial and error. Accordingly, we can think of collaboration as one element giving rise to "ICT dynamic capability" within knowledge communities.

Another important perspective is to promote deep dialectical dialog within knowledge communities through constructive approaches to ICT applications. A great deal of existing research (see, e.g., Duarte & Tennant Snyder, 2000; Jarvaenpaa & Leidner, 1999; Lipnack & Stamps, 1997; Odenwald, 1996; O'Hara-Devereaux & Johansen, 1994) tells us that human communication is generally the most important element in organizational management, including virtual teams activating ICT. Whether it takes place in real or virtual space, however, the most important behavior for actors is deep dialog. Face-to-face communication is not limited to real space. VIN tools that have advanced in recent years due to the kind of technological innovations in broadband and video processing mentioned in Chapter 1 are achieving face-to-face communication with a sense of presence in virtual space. VIN tools in a broadband environment enable deep dialog and promote collaboration among actors. Many examples exist of multipoint connection units (MCUs) that are already established in some companies for use with mobile videophones being useful for carrying out frequent meetings in virtual space and sharing information, contexts, and knowledge (Ohira, Kodama, & Yoshimoto, 2003).

As explained in Chapter 2, actors' trial-and-error processes give rise to creative realization among actors. These processes aim beyond the univocal strategic significance of companies' ICT activity promotion raising corporate efficiency and productivity, to seeking out the competitive environment and cultures of individual workplace organizations, and pursuing their natural markets and potential customer needs. This "creative realization" becomes the trigger for the creativity and imagination of new ideas and concepts to germinate among organizational actors. There it generates thoughts and actions among actors that pursue creativity as well as efficiency through ITC activity. This pursuit promotes dialectical dialog among actors within knowledge communities. Conversely, moreover, dialectical dialog among actors becomes the trigger for the generation of new "creative realization." In this way, ICT-related "productive interpretation" and "creative realization" come to promote deep collaboration and dialectical dialog among actors. This collaboration and dialog within knowledge communities in turn creates new ICT value among actors (see the "resonance of value" described in Chapter 8) and simultaneously builds mutual trust. This mutual trust spans actors' knowledge and organizational boundaries, and becomes an important element in generating collaboration in new knowledge communities. A large body of established research reports how the business performance of the virtual teams applying ICT relies on building mutual trust (see, e.g., Duarte & Tennant Snyder, 2000; Jarvaenpaa & Leidner, 1999; Lipnack & Stamps, 1997; Odenwald, 1996; O'Hara-Devereaux & Johansen, 1994; Zigurs, 2003).

Theories on trust are based on the notion of interdependence between the party who trusts and the party who is trusted (Dasgupta, 1988). Trust is an enabling condition that facilitates the formation of ongoing knowledge networks (Ring & Van den Ven, 1994), and some trust is required to initiate collaboration (Webb, 1991). Thus, deep dialog in collaboration helps to build trust because it provides the basis for continued interaction among knowledge communities, as described in the case studies in Chapters 4–8 (Leifer & Mills, 1996). From the analysis of these cases, it was found that trust could be gained through deep collaboration and dialectical dialog among the knowledge communities, where it serves as an important element in new knowledge creation relating to new business solutions and models for ICT applications (Bradach & Eccles, 1989; Gulati, 1995).

Meanwhile, according to the existing research relating to virtual teams up to now, reports have noted the difficulty of building mutual trust among actors in virtual teams where there is absolutely no real space, the "physical touch" of face-to-face communication is impossible (Handy, 1995), and the need for virtual team leaders to build trust in such ways as implementing suitable person-to-person meetings (Duarte & Tennant Snyder, 2000; Jarvaenpaa & Leidner, 1999; O'Hara-Devereaux & Johansen, 1994; Walker, Walker, & Schmitz, 2002). As mentioned above, however, advanced VIN tools technologically enable the sense of presence in face-to-face deep dialog, albeit in virtual space, and the viewpoint that it is difficult to build sufficient trust in the absence of real space will surely have to be reconsidered in the future. Realistically, a large number of the corporate actors I have surveyed apply VIN tools dynamically, and implement new strategies and tactics while creating new context and meaning through deep dialog in virtual space. Through the use of VIN tools, moreover, actors build mutual trust while sharing detailed nuances and feelings, and quickly make decisions on important business-related topics. When necessary, moreover, actors incorporate meetings in real space while producing new contexts and knowledge (real-space, face-to-face meetings may be deemed necessary when discussing workplace observations, discussing while touching products, joint experiments, and training requiring the transfer of high-level tacit knowledge, but future technological innovations will enable all of these cases to be carried out in virtual space).

The deep collaboration and dialectical dialog occurring in knowledge communities become elements that enhance the spiral "resonance of value" and the level of trust among actors. With knowledge communities that have built resonance of value, mutual trust among actors is enhanced, and sustained deep collaboration and dialectical dialog among actors builds new value and enables mutual trust in a spiral pattern (see Figure 9.1).

Meanwhile, the deep collaboration and dialectical dialog within knowledge communities helps actors realize the hidden contexts of knowledge communities, and produce the context architect capability that should give rise to further new context. Moreover, as mentioned in Chapter 8, the construction of the spiral resonance of value enhances ICT application capability and promotes dynamic ICT application. Building trust among actors, moreover, produces "boundaries consolidation capability" to form human networks spanning the boundaries of organizations and specialist knowledge. Then the ICT dynamic capability loop and community knowledge creating cycle (as shown in Chapter 7, Figure 7.5) are linked spirally as these actors' context architect, ICT application, and boundaries consolidation capabilities interact (see Figure 9.1).

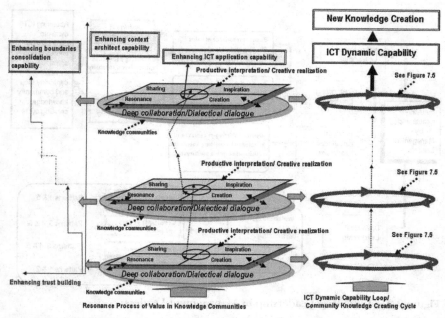

Figure 9.1. Deep Collaboration, Dialectical Dialogue, Resonance of Value, and Building Trust

This loop, however, does not end after a single revolution. Actors respond to environmental and contextual change by influencing the subjective and objective environments aimed at further change (or reproduction) through interaction with the enacted environment. Then the current ICT activity formation is improved, and further new ICT development and installations are promoted. With the processes that influence these actors' subjective environment, the building of new human networks by rebuilding existing knowledge communities in response to context (including the expansion of existing knowledge communities and network expansion) takes place at the same time (see the case study in Chapter 5), and new contexts and further ICT dynamic activities are promoted. As a result, actors create new context architect, ICT application, and boundaries consolidation capabilities, and the new ICT dynamic capability loop and community knowledge creating cycle are launched in a spiral pattern. Then new ICT dynamic capability and knowledge is produced as a result (see Figure 9.1).

Innovative Leadership by Management Leaders

In this section, I consider the ideal leadership that top and middle management leaders should undertake with regards to the process of acquiring

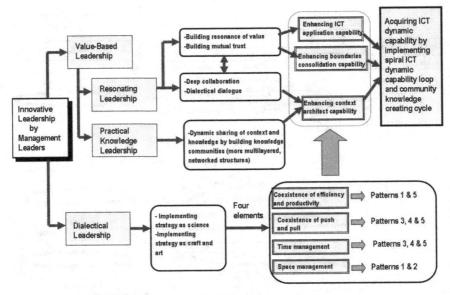

Figure 9.2.   Innovative Leadership by Management Leaders

organizational ICT dynamic capability from the introduction of ICT. In this book, I use the term "innovative leadership" for the leadership of the management leaders promoting new knowledge creation through ICT dynamic capability. I am presenting a number of core concepts from the multiple case studies detailed from Chapters 4–8, other examples of ICT development and introduction (52 cases), and the data that I have gathered on the subject of leadership through observations as a participant and ethnographist at the ICT development workplace. The first concept is "value-based leadership." This can be broken down into the two elements of "resonating leadership" and "practical knowledge leadership." The second concept is "dialectical leadership." This can be broken down into the four components of the dialectical management of strategy, power, time, and space (see Figure 9.2). I flesh out these concepts in the section below.

Value-Based Leadership

One element of "value-based leadership" is the management leaders' capacity to build resonance of value and trust. These aspects are produced from standardized coherence of individual interpretations with regard to strategic vision and goals for introducing ICT by playing off each other through a process of deep collaboration and dialectical dialog of actors'

mutual subjectivity and values. In other words, deep collaboration and dialectical dialog enable ICT strategy visions and goal sharing among actors.
Realistically speaking, however, interpretations of visions and goals differ
greatly depending on the actor concerned. This is where the management
leaders must repeatedly return to the starting points with the questions
"Why?" and "What are we doing this for?" and undertake a process of deep
collaboration and dialectical dialog among themselves. Then a sense of
unity is created among the actors over the formulation and implementation of ICT strategy, by means of mutual understanding of individual interpretations and the sharing and resonating of values. Next, the resonance of
value (Kodama, 2001a) and mutual trust mentioned in the section above
enhances actors' ICT application and boundaries consolidation capabilities. To build this resonance of value and mutual trust, management need
to promote leadership among actors. I call this "resonating leadership."

Another element of value-based leadership relates to the practical leadership ("What?" "How?") of management leaders and organization members aiming to create new value from ICT strategies through resonance of
value and mutual trust among management leaders. Value-based leadership shares high-quality, practical knowledge among actors (Boland & Tenkasi, 1995; Brown & Duguid, 1991, 1998; Cook & Brown, 1999; Hutchins,
1991, 1995; Lave, 1998; Orr, 1996; Schon, 1983, 1987; Spender, 1992; Suchman, 1987; Tsoukas, 1996; Weick & Browning, 1986; Wenger, 1998; Wenger
et al., 2002), and is an element of leadership as a process that actors apply.
This is a formation of the knowledge community including the community
of practice and strategic community mentioned in Chapter 2. Actors
exhibit leadership for the practically powerful context architect capability
that is a strength of actors aiming to create new contexts through deep collaboration and dialectical dialog within the knowledge communities. I call
this management leaders' "practical knowledge leadership."

This resonating and practical knowledge leadership produces the three
components of ICT dynamic capability (ICT application, boundaries consolidation, and context architect), initiates the ICT dynamic capability
loop and community knowledge creating cycle, and creates dynamic capability and new knowledge as an organization. "Value-based leadership" is a
general term with the meaning of these two leadership concepts creating
knowledge as new value. More specifically, "value-based leadership" components comprise management leaders forming resonance of value and
mutual trust in knowledge communities in order to implement individual,
specific ICT strategies and create new knowledge, and actors' capability in
discovering and implementing appropriate decisions and optimum action
from practical knowledge.

Dialectical Leadership

"Dialectical leadership" is another important leadership component in the acquisition of an organization's ICT dynamic capability. Dialectical leadership has two elements. One involves actors implementing strategy as analysis (science) to formulate and execute ICT strategy systematically and efficiently. The other involves actors implementing strategy as art and craft to produce and execute ICT strategy creatively, intuitively, or experientially. Dialectical leadership can be broken down into the four components of dialectical management of strategy (synthesis of efficiency and creativity), power (coexistence of push and pull), time, and space.

Dialectical management of strategy (syntheses of efficiency and creativity). One important element of the dialectical leadership that synthesizes the different behaviors of leadership's dominant dualities is that the management leaders in the knowledge communities (who use ICT dynamic capability) exhibit both strategic and creative leadership. Strategic leadership, based on integrated, centralized leadership, can create long- or short-term ICT strategy, focus on the big picture or pressing issues, and perform efficiently and reliably. Creative leadership, based on autonomous, decentralized leadership, can produce creative thinking and behavior among organizational members.

Top and middle management leaders must exhibit strategic leadership to efficiently, systematically, and reliably achieve short- and long-term ICT strategy goals relating to current mainstream business operations. These are the necessary components of leadership to promote the Pattern 1 strategy in Chapter 3 (Chapter 4's case studies also conform to Pattern 1). With deliberate strategy as the goal of short- and long-term ICT strategy, the management leaders in each division face the challenges of exhibiting strategic leadership and upsetting the status quo. Then, the management leaders' transformation of the organizations through incremental introduction of ICT becomes an important work function (Conger & Kanungo, 1998; Paul, Costley, Howell, & Dorfman, 2002; Politis, 2001; Popper & Lipshitz, 2000; Sadler, 2001). Furthermore, management leaders must produce new context and meaning related to current and future ICT strategies to subordinates, and bring about change in the subordinates' values, attitude, and behavior.

Meanwhile, management leaders must simultaneously maintain and exhibit the contrasting leadership components mentioned above. This is "creative leadership." By exhibiting creative leadership, actors must creatively and flexibly implement long- and short-term strategic goals regarding future ICT strategies. This is the leadership component essential to promoting Chapter 3's Pattern 2 strategy. Few management leaders in top management teams have practical experience of implementing Pattern 2 strategies (some energetic, entrepreneurial top management members brimming with imagination and new ideas do exist, however). Instead, it becomes the role of top management to lend support to middle management (or lower management team) leaders for this kind of creative ICT development environment (Pattern 2 is represented in Chapter 5's case studies). The kind of top management team that easily exhibits emergent or entrepreneurial strategy (Kodama, 2003b) based on creativity and imagination aimed at encouraging middle management to introduce and develop new ICT must build firm trust with the middle management team. Then top management constantly monitors future exploration activity, and manages essential resource distribution and the decision-making processes. Middle management leaders themselves must also display autonomous, decentralized leadership to subordinates and enhance subordinates' new concepts and creativity.

Thus management leaders must simultaneously demonstrate the contrasting elements of efficiency and creativity. Chapter 3, Pattern 5 implements such a strategy, and also corresponds to the management leadership of the ICT-driven, community-based firm in Chapter 8's case studies. The pursuit of efficiency requires the capability to discover problem-solving power or optimum solutions from an analytical or structural approach. Meanwhile, in the pursuit of creativity, it is important for management leaders to cultivate the power to imagine and predict, confirm provisionally, and grasp things that cannot be seen through an interpretive or process approach.

Dialectical management of power (push-and-pull synthesis). What kind of leadership must top and middle management leaders display within and outside the organization in order for a company to maintain competitive dominance from implementing ICT strategies? As Kotter (1999) suggests, leadership formulates visions and strategies, and concentrates actors' knowledge toward implementing these strategies. Leadership also empowers actors to realize their vision, and requires the power to overcome difficulties. Kotter (1982, 1988, 1990) emphasized the importance of the dual axes of "vision" and "network." To realize business as a new vision, it is important to form human networks (knowledge communities), including customers, within and outside the company. The vision structure (Bennis & Nanus, 1985; Kotter, 1988; Tichy & Devanna, 1986) stemming from this kind of transformational leadership is especially important at the top management level. Meanwhile, the human network structure provides a greater role for middle as well as top management.

Moreover, transformational leadership provides an insight different to that of the traditional leadership style, with its top-down component of executing work duties through commands to subordinates (requiring "pull" leadership, with the meaning of pulling subordinates upward). Transformational leadership empowers actors and motivates them to achieve their vision. Then management leaders inspire subordinates with new power (requiring "push" leadership, with the meaning of motivating subordinates by pushing them toward action) so that the subordinate actors carry out their work functions while demonstrating their leadership and collaborating with others.

Transformational leaders are constantly questioning current solutions, thinking differently, and encouraging creativity and innovation. They also relate to actors on a personal level, treating each as an individual with particular needs and abilities. This type of leadership inspires followers to transcend their own self-interest for the good of the organization (Burns, 1978; Paul et al., 2002; Politis, 2001; Popper & Lipshitz, 2000; Sadler, 2001).

Components that relate to management leaders' push-and-pull power balance and synthesis (acquired from case studies in Chapters 6, 7, and 8 in this book and in other field research) are important for dialectical leadership. Specifically, management leaders not only exhibit "forceful leadership" as directors who can take charge and control organizational members, but also become listeners, recipients, and collaborators based on "collaborative leadership" (Bryson & Crosby, 1992; Chrislip & Larson, 1994), empowering community members by enabling leadership and enhancing intrinsic motivation (Osterlof & Frey, 2000) among organizational members in their knowledge creation activities.

Put another way, the two contrasting components of leadership coexist, requiring the "forceful leadership" of a top-down leadership style to be a management leader, and the transformational leadership component to empower actors as knowledge workers. The actors' roles as supporters and followers, providing ongoing collaboration and support for the community so that it can pursue dreams and a sense of accomplishment for the business and its vision, requires the element of "servant leadership" (Greanleaf, 1979; Spears, 1995). Moreover, "collaborative leadership" involves an "orchestra with no conductor" (Seifter & Economy, 2001), with each actor displaying individual leadership and executing work functions in his or her area while implementing overall ICT strategy through collaboration with others. During this time, management leaders must act to maximize subordinates' capabilities as followers or servants.

Dialectical management of time. As discussed in the case studies for Chapters 6, 7, and 8, an important focal point for management leaders formulating and implementing ICT strategy is the "timing-specific" component of when the leaders will decide on and form the different strategy-making processes. Management leaders require leadership to implement patterns 3, 4, and 5 of the ICT strategy-making process. Realistically, this kind of leadership is required of top management in each department in a company, from the level of general supervisors and senior executive managers upward. The CEO, for example, must take a general view of the company with an integrated synergy that focuses on optimizing business domains both individually and as a whole, and must consider these with regard to two types of time-based strategies. Meanwhile, top management leaders who have been given responsibility through delegation of authority must sufficiently consider and formulate time-based ICT strategies for the business domains in their charge that optimize both the leaders' divisions and the company as a whole. An important aspect here is the kind of focus that management should give when formulating and implementing workplace ICT strategy.

As discussed in the case studies in Chapters 7 and 8, the implementation of top-down ICT strategies creates few cases of friction and conflict between head office and workplace organizations. However, the question of whether the importance of building top management's vision and achieving ICT strategy through storytelling can resonate with actors in the workplace organization is key to the success of the ICT strategy. This is why management leaders must demonstrate "resonating leadership," build resonance of value and mutual trust, and build knowledge communities spanning head office and workplace organizations. Building this resonance of value and mutual trust with the knowledge communities transforms conflict and friction to collaboration, and enables strategy-making processes to be managed dialectically on different time axes. In this way, close sharing and strong linkage of vision, ICT strategy, and organizational culture among actors in the head office and workplace organizations is essential to implementing the ICT strategy from the dialectical chronology of the strategy-making process.

Dialectical management of space. When management leaders are formulating and implementing ICT strategy, it is important to embed the connection- and network-specific components in the decision-making process. Management leaders aiming to create new knowledge face the question of how to build specific connections and networks amid constant change. As discussed in the case studies in Chapters 4 and 5, it is important that managers have the leadership to exploit network thinking (Kodama, 2007a, 2007b, 2007c) in ways that build knowledge communities, build more networked knowledge communities, and create new contexts by linking different organizational boundaries.

The building of tightly coupled, networked knowledge communities, for example, creates a strong value chain from a vertically integrated business model (Kodama, 2007a). With the conglomerate network of Chapter 4's automobile component manufacturers, for example (see Amasaka, 2004), deep knowledge sharing and integration was implemented as a result of the after-sales service of the automobile and component manufacturers through the building of tightly coupled, networked knowledge communities. In order to integrate knowledge in knowledge communities distributed inside and outside the company, actors closely integrate individual knowledge communities and embed diverse knowledge in networked knowledge communities through strong network ties.

Moreover, management leaders need to rebuild the networked knowledge communities as strong ties creating rigid value chains (Porter, 1985) and value networks (Christensen, 1997) as a result of environmental conditions and changes to markets and technologies. Put another way, management leaders convert networked knowledge communities with strong ties to loosely coupled networked knowledge communities, or in some cases, it may be necessary to show the leadership needed to terminate the knowledge communities (Kodama, 2007a. 2007b, 2007c). Otherwise, companies will face issues such as path dependency (Hargadon & Sutton, 1997; Rosenberg, 1982), competency traps (Levitt & March, 1988; Martines & Kambil, 1999), and core rigidities (Leonard-Barton, 1992, 1995).

Meanwhile, management leaders must form multiple knowledge communities within and outside the company in loosely coupled networked, knowledge communities, and experiment with peripheral region "scanning" and boundaries scanning among industries. The reason is that rapid ICT development has increased the need to merge and integrate different technologies spanning industry boundaries and to build diverse business models. Accordingly, management leaders must dynamically promote consortiums, incubations, joint experiments, and other methods with potential partners and customers outside the company spanning industry boundaries. At this point in time, however, knowledge communities within the company (internal knowledge communities, including top management, that influence decision making in the company) are united through weak network ties, and management leaders seek out and monitor future business opportunities through the ties among these "weak knowledge communities." Then the management leaders accurately sense the timing of the business opportunities, and convert the weak ties to strong ties at a stroke, thus achieving a new business model that realizes knowledge integration (Kodama, 2007b) (see Figure 9.3).

In Chapter 4's finance and Chapter 5's medical business cases, a business model was always required for advanced ICT, and management leaders had to implement this strategy of space. Put another way, it became important to dynamically create new business models crossing industry boundaries together with time-based change through the skillful management of the management leaders' tightly and loosely coupled, networked knowledge communities. Management leaders must respond to environ-

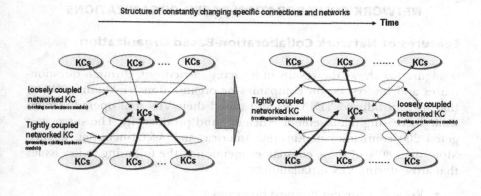

Figure 9.3.   Dialectical Management Realizing "Space Strategy"

mental change (or create their own new environments) by building ever-changing specific connections and networks with coexisting network ties having the contrasting characteristics of tightly and loosely coupled net-worked knowledge communities (see Figure 9.3). To realize this, manage-ment leaders must possess elements of "practical knowledge leadership" based on diverse experience and practical ability. Dialectical leadership is an important element for management leaders promoting this dialectical management of space.

Above, I described the elements of management leaders' "dialectical lead-ership." By exploiting this leadership, organizational members, including management leaders themselves, are able to participate in decision making in the knowledge communities and to enhance mutual understanding and strengthen links among the knowledge communities. Management leaders' dialectical leadership, which comprises four elements, enhances the three elements of ICT dynamic capability—ICT application, boundaries consolida-tion, and context architect—through a process of ICT strategy formulation and implementation (see Figure 9.2). Then, as a result of ICT dynamic capa-bility through dialectical leadership (dialectical management of strategy, power, time, and space), management leaders, including organizational members, can create new knowledge of business process reforms and the introduction and development of ICT. This image of leadership is not the old model buttressed by a rigid hierarchy but a new model of leadership aimed at achieving ICT innovation. This new type of leadership, dialectical leadership, is oriented toward the growth not just of individuals but of groups or organizations in the form of knowledge communities.

NETWORK COLLABORATION-BASED ORGANIZATIONS

Features of Network Collaboration-Based Organization

Figure 9.4 shows the results of a survey carried out, through question-naires and interviews, on companies or organizations (including NPOs) that had introduced VIN tools and raised their VIN tool operating ratio while increasing management efficiency and productivity. The survey targeted 200 companies in Europe, America, and Asia (including Japan). More than 80% of these companies mentioned the following as key issues that arose during VIN installation:

- Review of current business processes
- Review of decision-making processes (such as simplifying decision making)
- Promoting delegation of authority
- Flattened organizational structure
- Staff training during ICT introduction
- Promotion of knowledge management

The companies and organizations where these six issues frequently appeared are named "network collaboration-based organizations" in this book. I believe that future corporate organizations' styles will migrate to

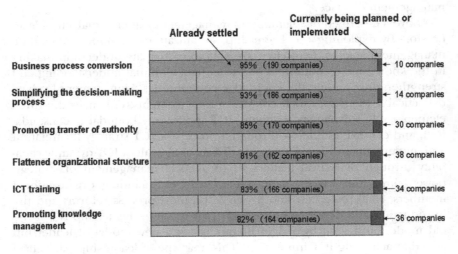

Source: Results of survey from 2004 to 2006

Figure 9.4. Features of Network Collaboration-Based Organizations

those of network collaboration-based organizations. The survey identified the following shared features of these organizations:

- Promotion of collaboration in real and virtual space
- High level of communication and information sharing within and among organizations promoting knowledge management
- Presence of activated knowledge communities
- Establishing advanced delegation of authority and rapid decision making
- Flat organizational structure
- Strong focus on results

The following three points were also mentioned as new behavior among organization members introducing VIN tools. First was a rise in informal companies, and the creation of new contexts as a result. Second was a rise in pragmatic knowledge communities aimed at solving issues, especially an increase of actors' voicing of opinions in virtual space. Third was the promotion of new time and space management. Next, I mention the time and space strategies of network collaboration-based organizations with regards to the promotion of this new management.

Implementing Strategy in Time and Space through ICT

A major feature of the network collaboration-based organization is the implementation of dialectical strategy in time and space. Actors in network collaboration-based organizations exploit their ICT application capability to promote collaboration by interacting in real and virtual space as a space strategy. As shown in the new product development case in Chapter 2, Figure 2.4, actors in real and virtual space exploit context architect capability through dialog and practice to share contexts while converting to more dynamic contexts. New contexts are created as a result, and linked to actors' new action. Actors in virtual space apply VIN tools while sharing information and contexts on demand, in real time. Then management leaders form new knowledge communities in real and virtual space through boundaries consolidation capability among organizations. This is a component of network collaboration management on a space axis (see Figure 9.5).

Meanwhile, as a time strategy, network collaboration-based organizations promote collaboration through interacting on two different time axes. One, called "planned time," is a time that is already fixed objectively as clock time. Planned time mainly involves the analytical and rational pro-

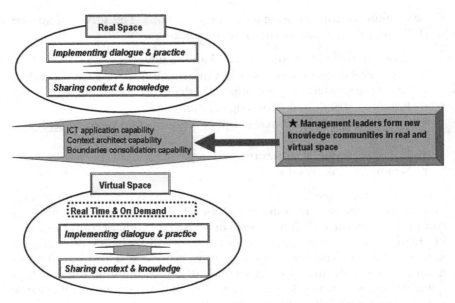

Figure 9.5.   Network Collaboration Management on a Space Axis

cesses of organizations' decision-making and strategy-making processes. The existing business operations of already established companies (including routine product development and production techniques) are implemented using planned time. The main topics of planned time strategy goals are the pursuit of efficiency, improvement activities, and best practice. Moreover, the nature of a network collaboration-based organization with regard to planned time is that actors' activities mostly involve routine work functions through collaboration as an official organization, and official conferences activating VIN tools are implemented on planned time axes. With these axes, actors enhance ICT application capability through day-to-day VIN tool activity.

The second time axis is called "emergent time," a subjective "timely time" that corresponds to time subjectively redefined by actors. With emergent time, an organization's decision-making and strategy-making processes possess the elements of improvised processes, and require actors to make snap decisions. As metaphors, planned time corresponds to an orchestra, while emergent time corresponds to jazz or the theater. Strategy goals with emergent time mainly comprise the pursuit of creativity and the creation of new business and innovations. The nature of network collaboration-based organizations using emergent time is one where the actors' main task is to collaborate through informal networks as informal knowledge communities (including temporary projects, CFT, and task teams) as a result of actors' boundaries consolidation capability. Moreover, the creation of themes and problem solving as a result of temporary meetings, using VIN tools on this emergent time axis, demonstrates that actors have accumulated ICT application capability day by day on a planned time axis. With this emergent time, the skills required of actors are creativity, imagination, and flexibility, which are different from the logical thinking required of planned time. It is important for management leaders to create new contexts from the two differentiated types of time management (coexisting and separated by use) through context architect capability. The paradoxical management of these two time axes is a component of network collaboration management on a time axis (see Figure 9.6).

Chapter 3 contains a case where Pattern 1 implements strategies on a planned time axis, and Pattern 2 implements strategies on an emergent time axis. As regards this, Patterns 3 and 4 include cases that cleverly combine planned and emergent time while redefining, recreating, and then formulating and implementing strategies on a planned time axis as a trigger for emergent time. Pattern 5 also simultaneously combines planned and emergent time, and redefines and recreates strategy on a planned time axis by strategically (deliberately) connecting these time axes (see Figure 9.7).

Planned time				**Emergent time**	
Time quality	Objectively set time (clock time)		Time quality	Subjective time redefined by self (timely time)
Decision-making and strategy formation (metaphor)	Analytical and rational processes (orchestra)	ICT application capability	Decision-making and strategy formation (metaphor)	Improvised processes and instant decisions (jazz and the theater)
Strategic goals	- Pursuit of efficiency - Enhancing activities - Best practice	Context architect capability	Strategic goals	- Pursuit of creativity - Creating a new vision - Innovation
Qualities as network collaboration-based organization	Executing duties through collaboration as formal organization	Boundaries consolidation capability	Qualities as network collaboration-based organization	- Informal, temporary organization (temporary projects, task teams etc.) - Collaboration through informal networks
Required skills	Logical thinking		Required skills	Creativity and imagination

★ It is important for management leaders to create new knowledge from paradoxical management of two different times (allowing coexistence and differentiated use)

Figure 9.6.   Network Collaboration Management on a Time Axis

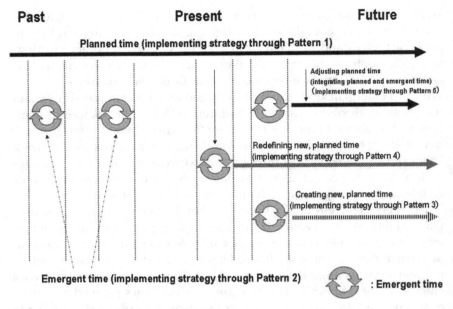

Figure 9.7. ICT Strategy on a Time Axis

Network Collaboration-Based Organizations and Networked Knowledge Communities

The basic structural components of network collaboration-based organizations are the knowledge and networked knowledge communities created from actors at flat-structured formal and other formal organizations (see Figure 9.9). The formation of knowledge communities enables interaction of dialog and practice among individuals and an organizational body that creates new knowledge. As shown in Chapter 7, Figure 7.5, knowledge communities are platforms that become media for dynamically linking the ICT dynamic capability loop and the community knowledge creating cycle, and are also media and catalysts to actually drive the knowledge creation process.

Management leaders (leaders who share visions with top leaders and promote the knowledge creation process at the workplace) create ICT dynamic capability and new knowledge by forming knowledge communities from the innovative leadership mentioned above, and also, where necessary, through interaction with knowledge communities and with customers, external partners, and others. Knowledge is closely connected with situation, scene, and space—in other words, with context. Actors need to share context in order to acquire, share, and create knowledge, yet they cannot share unless they participate in a specific knowledge community with a specific context, space, and time. The medium of shared context enables a link with the community knowledge creating cycle, comprising the sharing, inspiring, creation, and accumulation of knowledge. At this time, hidden context is revealed through the management leaders' context architect capability. Actors share this context and go on to form new contexts.

To create new knowledge from still higher quality ICT dynamic capability, management leaders participate in multiple, differentiated knowledge communities, circulate context and knowledge among these communities, and act to share and inspire this knowledge among the community actors. This demonstrates the boundaries consolidation capability of the management leaders, who implement the creation of networks in multiple, differentiated knowledge communities. Furthermore, the actors share context and knowledge in virtual space, which complements the ICT activities, and at the same time enhance ICT application capability (see Figure 9.8). Then the management leaders create new contexts for the networked knowledge communities, and further enhance context architect capability.

These knowledge communities and formal organizations are not opposites. Rather, the knowledge communities embrace management methods

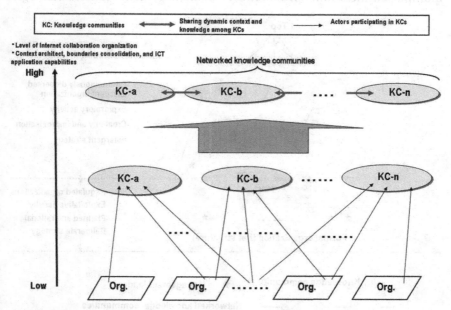

Figure 9.8.   Networked Knowledge Communities

centered on problem-solving tasks from the routine and daily task management of formal organizations. Management leaders and organization members not only observe the basic rules and decision-making processes of formal organizations, but also manage the self-distributive, challenging organizational behavior at knowledge communities, and at the same time coordinate, expand, and develop knowledge communities. Considering organizational management from the standpoint of knowledge communities, liberation from the management knowledge thinking of the formal organization brings to light completely new management models. Put another way, depending on context and environment, management leaders in formal organizations create the structure of knowledge communities spanning formal organizations, and their networked structures (rebuilding knowledge communities), and flexibly transform the knowledge communities in response to strategic goals.

The network collaboration-based organization part of the formal organization does not mean a hierarchical, bureaucratic organization (with the image of a mechanized organization) with an unchanging layered structure. Instead, as described above, it is a flat-structured (two to four layers) formal organization establishing faster delegation of authority and decision making (see Figure 9.9). This flat formal organization is formed from multiple business units presided over by management leaders. The busi-

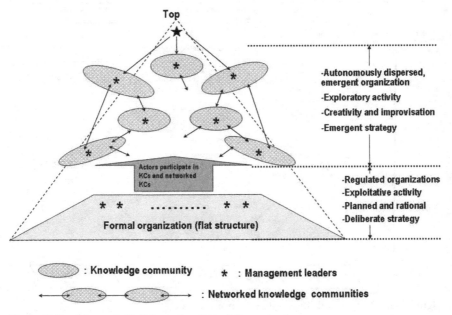

Figure 9.9. Network Collaboration-Based Organizations

ness unit managers and management leaders collaborate with those of other business units, depending on context, and form self-distributive knowledge communities. External partners and specific customers are also embedded in the knowledge communities as required.

Knowledge communities need to fragment and merge as the complexity of business models and issues continues to grow. Put another way, actors' formation of networked knowledge communities enables the integration of knowledge produced by different knowledge communities distributed within and outside the company (Kodama, 2007a, 2007b, 2007c). Knowledge communities may comprise organizational structures possessing a targeted objective or issue-solving mission, and also include the project and cross-functional team (CFT) organization structure concepts. The organizational behavior where multiple projects build a projects network (Kodama 2007c) and different project members collaborate while building new business models corresponds to the networked knowledge communities. The more networked the knowledge communities, the higher their level as a network collaboration-based organization (see Figure 9.8). With time management of these organizations, actors implement routine work functions and improvements on a planned time axis with formal organizations, while enjoying the challenge of activating high-quality (or highly difficult) business on an emergent time axis with knowledge or networked knowledge communities. Managing this kind of network collaboration-based organization structure requires management leaders to demonstrate a new management style with innovative leadership (as mentioned above).

Building Network Collaboration-Based Organizations

When realizing a network collaboration-based organization, new knowledge cannot be produced simply by introducing ICT such as VIN tools, and having actors apply them. To take the example of Buckman Laboratory, a cutting-edge company that applies ICT and promotes knowledge management (Buckman Laboratory employs 1,300 staff in research laboratories and business offices in 21 countries and undertakes research and development from paper manufacturing companies, among others), CEO Buckman says the following.

> This does help, at least initially, by causing knowledge that has been written down to be organized and made available to the organization. Useful as that step is, however, we've found that it is not sufficient to achieve success. It can deal with no more than a small fraction of knowledge in the company—perhaps about 10% of the total. The dynamics of a company don't change when it organizes its explicit knowledge.

We found that the vast bulk of the knowledge in Buckman Labs was in the heads of our people—and it was changing every minute of every day. It was not written down. Therefore, if we wanted to achieve success in the fast-changing environment confronting us, we had to learn how to engage people and arouse their interest and trust, making them willing to move their knowledge across the organization to where it was needed, when it was needed. (2003, p. 20).

As mentioned in Chapter 2, ICT in itself basically promotes sharing of explicit knowledge, but ICT alone will not give a company a competitive edge. Company executives and management leaders must arouse actors' interest in business themes, share trust and value among the actors, and build knowledge communities. Then it is important for the actors in the knowledge communities to promote the knowledge sharing and inspiring process while dynamically applying ICT, and create and accumulate new knowledge.

A characteristic feature of Buckman Laboratory is its employees' unique working style. When new product concepts are dreamed up but research reaches the end of the road, the employees face the PC and, without fail, access one particular page: an Internet forum by the name of "Techforum." Techforum is a web-based forum for knowledge sharing, inspiration, creation, and accumulation (the community knowledge creating cycle). Each country's employees access Techforum and check to see whether any other employees with the same kind of knowledge and expertise are around. If a large number of employees transcending national boundaries, specializations, sections, and organizations agree, a project can be started up independently on Techforum. The employees bring their knowledge together and discuss issues while advancing research and development work on successively emerging projects on the Internet. More than 20 new technologies are created each year from this kind of project, especially in paper manufacturing techniques, which is an area of special expertise. Through the knowledge creation process of the Techforum knowledge communities, individual employees talk and think dynamically to share tacit knowledge, inspire each other, and endlessly create the new knowledge of new product development.

In cases such as Buckman Laboratory's, the creation of new knowledge (innovation) has a strong tendency to create boundaries among disciplines and specializations (Leonard-Barton, 1995). The company is divided into various work organizations and specializations where a large number of boundaries, both visible and invisible, exist (these might include geographic boundaries as globalization, industry boundaries as strategy, organizational boundaries as theory of the firm, and human cognition as bounded rationality). In an industry experiencing a changing environment and fierce competition, many companies are transcending numerous boundaries inside and outside the company (including customers), and actors are integrating diverse knowledge and implementing strategy with the aim of innovating and creating new knowledge. Knowledge is certainly a source of corporate competitive power (Kogut & Zander, 1992; Leonard-Barton, 1995; Nonaka & Takeuchi, 1995), but the organizational boundaries of sectionalism among individual corporate actors and the knowledge boundaries arising from each actor's values, background, and specializations (Brown & Duguid, 2001) also exist. Thus actors' inherent mental models (Grinyer & McKiernan, 1994; Spender, 1990) and path-dependent knowledge can become a barrier to innovation (Carlile, 2002).

So how can organizational and knowledge boundaries be surmounted, actors' knowledge managed, and new knowledge be created? In order to do this, it is important to share missions among actors, and to have a means of inducing productive interpretation and creative realization in actors with regard to business. Then, resonance of value and mutual trust must be built through dialectical dialog and deep collaboration among actors, and knowledge (and networked knowledge) community structures promoted. Actors also need the space-management strategies to create new contexts and knowledge through dynamic interaction in real and virtual space, as shown in Figure 9.5. Here it is essential to go beyond a deep understanding of ICT among actors, for the company as a cooperative organization to cultivate a sense of unity among actors, and further, for individual actors' to understand each other's thinking and values.

The above-mentioned knowledge and networked knowledge communities are not opposed to formal organizations. There are two types of network collaboration-based organizations, and these two bodies complement each other (see Figure 9.9). Actors participating in knowledge communities (including networked knowledge communities) promote exploitative activities aimed at current business on a planned time axis for the formal organizations mentioned above, and at the same time pursue the exploratory activity of new business activating knowledge communities (including networked knowledge communities) on an emergent time axis.

Self-distributive knowledge communities (including networked knowledge communities) possess both the passive aspect of responding to environmental change and the active aspect of environment building by taking on the challenge of untapped markets and new business. Moreover, tasks at flat-layered, formal organizations are important components not just in executing existing business but also in implementing new business that arise from knowledge communities (including networked knowledge communities) rationally and systematically on an emergent time axis. It follows that the paradoxical management of the knowledge communities (including networked knowledge communities) comprising flat-layered organizations (image of regulated organizations) as formal organizations emphasizing principles and rational, deliberate strategy, and the emergent organizations (image of organic organizations) (Kodama, 2007a, 2007b, 2007c) emphasizing self-distributive, creative, and improvised behavior is important to the network collaboration-based organization.

Important points about the network collaboration-based organization are to keep building appropriate flat, formal organizations to optimize the company as a whole from the viewpoint of a company's new knowledge creation and new customer value, and at the same time always to rebuild dynamic knowledge communities and networked knowledge communities. Management leaders must be the architects of the knowledge communities (Kodama, 2007a, 2007b, 2007c), and this architect capability is a condition to keep innovating for a company's sustained competitive edge.

REFERENCES

AERA. (1999). *Gendai no Shozo* [in Japanese], *31*, 62–66.

Amasaka, K. (2004). Development of "Science TQM", a New Principle of Quality Management: Effectiveness of Strategic Stratified Task Team at Toyota. *International Journal of Production Research, 42*(17), 3691–3706.

Anand, B., & Khanna, T. (2000). Do Firms Learn to Create Value?: The Case of Alliances. *Strategic Management Journal, 21*(3), 295–315.

Anthony, R. N. (1965). *Planning and Control Systems: A Framework for Analysis. Division of Research.* Boston: Graduate School of Harvard University.

Argote, L. (1999). *Organizational Learning: Creating, Retaining and Transferring Knowledge.* Norwell, MA: Kluwer,

Barley, S. R. (1986). Technology as an Occasion for Structuring' Evidence from Observations of CT Scanners and the Social Order of Radiology Departments. *Administrative Science Quarterly, 31*, 78–108.

Barney, J. B., Wright, M., & Ketchen, Jr., D. J. (2001). The Resource-Based View of the Firm: Ten Years after 1991. *Journal of Management, 27*, 625–641.

Bechard, R., Goldsmith, M., & Fesselbein, F. (1996). *The Leader of the Future.* San Francisco: Jossey-Bass.

Bennett, R. (2001). Ba" as a Determinant of Salesforce Effectiveness: An Empirical Assessment of the Applicability of the Nonaka–Takeuchi model to the Management of the Selling Function. *Marketing Intelligence and Planning, 19*(3), 188–199.

Bennis, W. G., & Nanus, B. (1985). *Leaders: The Strategies for Taking Charge.* London: HarperCollins.

Benson, J. (1977). Organization: A Dialectical View. *Administrative Science Quarterly, 22*, 221–242.

Bharadwai, A. S. (2000). A Resource-Based Perspective on Information Technology Capability and Firm Performance: An Empirical Investigation. *MIS Quarterly, 24*(1), 169–196.

Boland J., & Tenkasi, R. (1995). Perspective Making and Perspective Taking: Implications for Organizational Learning. *Organization Science, 9*(3), 605–622.

New Knowledge Creation Through ICT Dynamic Capability, pages 197–207
Copyright © 2008 by Information Age Publishing

Boudreau, M.-C., & Robey, D. (2005). Enacting Integrated Information Technology: A Human Agency Perspective. *Organization Science, 16*(1), 3–18.

Bradach, J., & Eccles, R. (1989). Price, Authority, and Trust: From Ideal Types to Plural Forms. In W. Sewell (Ed.), *Annual Review of Sociology* (Vol. 15, pp. 97–118). Palo Alto, CA: Annual Reviews.

Brown, J. S., & Duguid, P. (1991). Organizational Learning and Communities-of-Practice. *Organization Science, 2*(3), 40–57.

Brown, J. S., & Duguid, P. (1998). Organizing knowledge. *California Management Review, 40*(3), 90–111.

Brown, J. S., & Duguid, P. (2001). Knowledge and Organization: A Social–Practice Perspective. *Organization Science, 12*(6), 198–213.

Brynjolfsson, E. (2000). *Understanding the Digital Economy: Data, Tools, and Research.* Boston: MIT Press.

Brynjolfsson, E., & Hitt, L. (1995). Computers as a Factor of Production: The Role of Differences Among Firms. *Economics of Innovation and New Technology, 2,* 183–199.

Brynjolfsson, E., & Hitt, L. (1996). Paradox Lost?: Firm-Level Evidence of High Returns to Information Systems Spending. *Management Science, 42*(4), 541–558.

Bryson, J., & Crosby, B. C. (1992). *Leadership for the Common Good: Tackling Public Problems in a Shared-Power World.* San Francisco: Jossey-Bass.

Buckman, R. (2003). *Building a Knowledge-Driven Organization,* New York: McGraw-Hill.

Burke, R. J. (1970). Methods of Resolving Superior-subordinate Differences and Disagreements. *Organizational Behavior and Human Preformance, 5.*

Burns, M. (1978). *Leadership.* London: Harper & Row.

Camerer, C., Lowenstein, G., & Weber, M. (1989). The course of knowledge in economic settings: An empirical analysis. *Journal Political Economics, 97,* 1232–1254.

Carlile, P. (2002). A Pragmatic View of Knowledge and Boundaries: Boundary Objects in New Product Development. *Organization Science, 13*(4), 442–455.

Carlile, P. (2004). Transferring, Translating, and Transforming: An Integrative Framework for Managing Knowledge Across Boundaries. *Organization Science, 15*(5), 555–568.

Carlile, P. R., & Rebentisch, E. S. (2003). Into the black box: the knowledge transformation cycle, *Management Science, 49*(9), 1180–1195.

Carlson, J. R., & Zumd, R. W. (1999). Channel expansion theory and the experiential nature of media richness perceptions. *Academy of Management Review, 42*(2), 153–170.

Carr, N.G. (2003, May). IT Desen't Matter. *Harvard Business Review,* pp. 41–49.

Carr, N.G. (2004). *Dose IT Matter?* Boston: Harvard Business School Press.

Chesbrough, H. (2003). *Open Innovation.* Boston: Harvard Business School Press.

Chrislip, D., & Larson, C. (1994). *Collaborating Leadership: How Citizens and Civic Leaders Can Make a Difference.* San Francisco: Jossey-Bass.

Christensen, C. M. (1997). *The Innovator's Dilemma: When New Technologies Cause Great Firms to Fail.* Boston: Harvard Business School Press.

Clark, K., Bruce, W., & Fujimoto, T.(1987). *Product Development in the World Auto Industry.* (Harvard University Brookings Papers on Economic Activity, 3). Boston: Harvard University Press.

Clemons, M. K. (1991a). Corporate Strategies for Information Technology: A Resource-Based Approach. *Computer, 24*(11), 23–32.

Clemons, M. K. (1991b). Evaluation of strategic investments in information technology. *Communications of the ACM, 34(1), 22–36.*

Clemons, M. K., & Row, M. C. (1991). Sustaining IT Advantage: The Role of Structural Differences. *MIS Quarterly, 15*(3), 275–292.

Conger, A., & Kanungo, R. (1998). *Charismatic Leadership: The Elusive Factor in Organizational Effectiveness.* San Francisco: Jossey-Bass.

Cook, S., & Brown. J. (1999). Bridging Epistemologies: The Generative Dance between Organizational Knowledge and Organizational Knowing. *Organization Science, 10*(2), 381–400.

Cramton, C. (2001). The Mutual Knowledge Problem. *Organization Science, 12,* 346–371.

Daft, R. L., & Lengel, R. H. (1984). Information richness: A new approach to managerial behavior and organizational design. In B. M. Staw & L. L. Cumming (Eds.), *Research in Organizational Behavior* (pp. 191–233). Greenwich, CT: JAI Press.

Daft, R. L., & Lengel, R. H. (1986). Organizational Information Requirements Media Richness and Structural Design. *Management Science,* 32(5), 554-571.

Daft, R.L., & Macintosh, N.B. (1981). A Tentative Exploration into the Amount and Equivocality of Information Processing in Organizational Work Units. *Administrative Science Quarterly, 26,* 207–224.

Daft, R. L., & Weick, K. E. (1984). Toward a Model of Organizations as Interpretation Systems. *Academy of Management Review, 10*(4), 803–813.

Damanpour, F. (1991). Organizational Innovation: A Meta-Analysis of Effects of Determinants and Moderators. *Academy of Management Journal, 34*(3), 555–590.

Dasgupta, P. (1988). Trust as a Commodity. In D. Gambetta (Ed.), *Trust: Making and Breaking Cooperative Relations* (pp. 49–72). New York: Blackwell.

Davenport, T. H. (1993). *Process Innovations—Reengineering Work through Information Technology.* Boston: Harvard Business School Press,

Davenport, T. H., & Prusak, L. (1998).*Working Knowledge.* Boston: Harvard Business School Press.

Davidow, W. H., & Malone, M. S. (1992). *The Virtual Corporation: structuring and revitalizing the Corporation for the 21st Century.* New York: HarperCollins.

DeSanctice, G., & Poole, P.(1994). Capturing the Complexity in Advanced Technology Use: Adaptive Structuration Theory. *Organization Science, 5*(2), 121–147.

Dougherty, D. (1992). Interpretive Barriers to Successful Product Innovation in Large Firms. *Organization Science, 3*(2), 179–202.

Duarte, D., & Tennant Snyder, N. (2000). *Mastering Virtual Teams: Strategies, Tools, and Techniques that Succeed.* San Francisco: Jossey-Bass.

Dyer, J. (1996). How Chrysler Created an American Keiretsu. *Harvard Business Review, 74,* 42–51.

Emirbayer, M., & Mische, A. (1998). What is agency?. *American Journal of Sociology, 103,* 962–1023.

Engels, F. (1952). *Dialektik der Natur.* Berlin: Dietz.

Ernst, M., & Young, J.(1991). *The Car Company of the future.* Ann Arbor: University of Michigan.

Evans, P., & Wurster, T. (1997). Strategy and the New Economy of Information. *Harvard Business Review, 75*(5), 71–82.

Ford, M., Lew, H., Spanier, S., & Stevenson, T. (1997). *Internetworking Technologies Handbook.* Indianapolis, IN: New Riders Publishing.

Fuller, S. (2001). *Knowledge Management Foundations.* London: Butterworth-Heinemann.

Giddens, A. (1979). *Central problems in social theory: Action, Structure and Contradiction in Social Analysis.* Berkeley: University of California Press.

Giddens, A. (1984). *The Constitution of Society.* Berkeley: University of California Press.

Grant, R. M. (1991, Spring). Resource-Based Theory of Competitive Advantage: Implications for Strategy Formulation. *California Management Review,* pp. 114–135.

Grant, R., & Baden-Fuller, C. (2004). A Knowledge Accessing Theory of Strategic Alliance. *Journal of Management Studies, 41*(1), 61–84.

Gray, B. (1989). *Collaborating: Finding Common Ground for Multiparty Problems.* San Francisco: Jossey-Bass.

Gray, B., & Wood, D. J. (1991). Collaborative Alliances: Moving from Practice to Theory. *Journal of Applied Behavioral Science, 27*(1), 3–23.

Greenleaf, R. (1979). *Servant Leadership.* New York: Paulist Press.

Grinyer, P., & McKiernan, P. (1994). Triggering Major and Sustained Changes in Stagnating Companies. In H. Daems & H. Thomas (Eds.), *Strategic Groups, Strategic Moves and Performance* (pp. 173–195). New York: Pergamon.

Gulati, R. (1995). Does Familiarity Breed Trust?: The Implications of Repeated Ties for Contractual Choice in Alliances. *Academy of Management Journal, 38*(1), 85–112.

Gulati, R. (1999). Network Location and Learning: The Influence of Network Resources and Firm Capabilities on Alliance Formation. *Strategic Management Journal, 20*(5), 397–420.

Hamel, G., & Prahalad, C. K. (1994). *Competing for the Future.* Boston: Harvard Business School Press.

Hammer, M., & Champy, J. (1993). *Reengineering the Corporation: A Manifesto for Business Revolution.* New York: HarperBusines

Handy, C. (1995). *Trust and the Virtual Organizations.* Harmondsworth, UK: Penguin.

Hargadon, A., & Sutton, R. (1997). Technology brokering and innovation in a product development firm. *Administrative Science Quarterly, 42,* 716–714.

Hegel, G. W. F. (1967). *The phenomenology of mind.* New York, Harper & Row.

Helper, S., & Sako, M. (1995). Supplier Relations in Japan and the US: Are they Converging?. *Sloan Management Review, 36,* 77–83.

Hesselbein, F., Goldsmith, M., Beckhard, R., & Schbert, R. F. (1998). *The Community of the Future.* San Francisco: Jossey-Bass.

Hinds, P. (1999). The Course of Expertise: The Effects of Expertise and Debiasing Methods on Prediction of Novice-Performance. *Journal of Experimental Psychology: Applied, 5,* 205-221

Hinds, P., & Kiesler, S. (1995). Communication across Boundaries: Work, Structure, and Use of Communication Technologies in a Large Organization. *Organization Science, 6*(4), 373–393.

Hirose, T. (1985). Veterinary Medicine: Results and Research Prospects. *Japan Veterinary Association*, pp. 233–237.

Hirose, T. (2000, August) Experience in using an integrated distant education by ISDN system in Japan. Paper presented at the 12th Meeting of the International Veterinary Radiology Association.

Hobday, M. (1998). Product complexity, innovation and industrial organization. *Research Policy, 26*, 689–710.

Hobday, M. (2000). The Project-Based Organisation: An Ideal Form for Managing Complex Products and Systems?. *Research Policy, 29*, 871–893.

Huber, G. (1991). Organizational Learning: The Contributing Processes and the Literature. *Organization Science, 2*(1), 88–115.

Hutchins, E. (1991).Organizing Work by Adaptation. *Organization Science,* 2(1), 14–39.

Hutchins, E. (1995).*Cognition in the Wild.* Cambridge, MA: MIT Press.

Jantsch, E. (1980). *The Self-Organizing Universe.* Oxford, UK: Pergamon Press.

Jarvaenpaa, S., & Leidner, D. (1999). Communications and trust in Global Virtual Teams. *Organization Science, 10*(6), 791–815.

Johansson, F. (2004). *The Medici Effect.* Boston: Harvard Business School Press.

Jones, M. (1999). Information Systems and the Double Mangle: Steering a Course between the Scylla of Embedded Structure and the Charybdis of Strong Symmetry. In T. J. Larsen, L. Levine, & J. I. DeGross (Eds.), *Information Systems: Current Issues and Future Changes* (pp. 287–302). Laxenburg, Austria: IFIP.

Jonscher, C. (1994). An Economic Study of the Information Technology Revolution. In T. J. Allen & M. S. Morton (Eds.), *Information Technology and the Corporation of the 1990s: Research Studies* (pp. 5–42). Oxford, UK: Oxford University Press.

Kale, P., Singh, H., & Perlmutter, H. (2000).Learning and Protection of Proprietary Assets in Strategic Alliances: Building Relational Capital. *Strategic Management Journal, 21*(3), 217–237.

Kanter, R., Kao, J., & Wiersema, F.(1997) *Breakthrough Thinking at 3M, Dupont, GE, Pfizer, and Rubbermaid.* London: HarperCollins.

Khurana, A., & Rosenthal, S. R. (1998). Toeards Holistic "Front Ends" in New Product Development. *Journal of Product Innovation Management. 15*(1), 57–74.

Kodama, M. (1999a). Community Management Support through Community-based Information Network. *Information Management and Computer Security, 7*(3).

Kodama, M. (1999b). Strategic Business Applications and New Virtual Knowledge-Based Businesses through Community-Based Information Networks. *Information Management and Computer Security, 7*(4).

Kodama, M. (2000). Business Innovation through Customer-Value Creation: Case Study of a Virtual Education Business in Japan. *Journal of Management Development, 19*(1), 49–70.

Kodama, M. (2001a). Creating New Business Through Strategic Community Management. *International Journal of Human Resource Management, 11*(6), 1062–1084.

Kodama, M. (2001b). Distance Learning Using Video Terminals: An Empirical Study. *International Journal of Information Management, 21*(3), 227–243.

Kodama, M. (2001c). New Regional Community Creation, Medical and Educational Applications through Video-Based Information Networks. *Systems Research and Behavioral Science, 18*(3), 225–240.

Kodama, M. (2002a)..The Promotion of Strategic Community Management Utilizing Video-based Information Networks. *Business Process Management Journal, 8*(5).

Kodama, M. (2002b). Strategic Partnership with Innovative Customers: A Japanese Case Study. *Information Systems Management, 19*(2), 31–52.

Kodama, M. (2002c). Transforming an Old Economy Company through Strategic Communities. *Long Range Planning, 35*(4), 349–365.

Kodama, M. (2003a). Strategic Community-Based Theory of the Firms: Case Study of NTT DoCoMo. *Journal of High Technology Management Research, 14*(2), 307–330.

Kodama, M. (2003b). Strategic Innovation in Traditional Big Business. *Organization Studies, 24*(2), 235–268.

Kodama, M. (2004). Strategic Community-Based Theory of Firms: Case Study of Dialectical Management at NTT DoCoMo. *Systems Research and Behavioral Science, 21*(6), 603–634.

Kodama, M. (2005a). Knowledge Creation through Networked Strategic Communities: Case Studies in New Product Development. *Long Range Planning, 38*(1), 27–49.

Kodama, M. (2005b). New Knowledge Creation through Dialectical Leadership: A Case of IT and Multimedia Business in Japan. *European Journal of Innovation Management, 8*(1), 31–55.

Kodama, M. (2006). Strategic Community: Foundation of Knowledge Creation. *Research-Technology Management, 49*(5), 49–58.

Kodama, M. (2007a). *The Strategic Community-Based Firm.* London: Palgrave Macmillan.

Kodama, M. (2007b). *Knowledge Innovation—Strategic Management As Practice.* London: Edward Elgar.

Kodama, M. (2007c). *Project-Based Organization in the Knowledge-Based Society.* London: Imperial College Press.

Kodama, M, Ohira, H., Kawakami, T., Kaneko, A., & Suzuki, T. (2004). Empirical Trials and Evaluation of an Industry/Academia Collaborative Agent System Supporting the Launch of New Enterprises. *International Journal of Management and Enterprise Development, 1*(2).

Kogut, B., & Zander, U. (1992). Knowledge of the Firm: Combinative Capabilities and the Replication of Technology. *Organization Science, 5*(2), 383–397.

Kotter, J. (1982). *The General Manager.* New York: Free Press.

Kotter, J. (1988). *The Leadership Factor.* New York: Free Press.

Kotter, J. (1990). *A Force for Change: How Leadership Differs from Management.* New York: Free Press.

Kotter, J. (1999). *Kotter on What Leaders Really Do*. Boston: Harvard Business School Press.

Kydd, C. T., & Ferry, D. L. (1994). Managing Use of Video Conferencing. *Information and Management, 27*(6), 369–375.

Larsson, R., Bengtsson, L., Henriksson, K., & Sparks, J. (1998). The Interorganizational Learning Dilemma: Collective Knowledge Development in Strategic Alliances. *Organization Science, 9*(3), 285–305.

Lave, J. (1998). *Cognition in practice*. Cambridge, UK: Cambridge University Press.

Lave, J., & Wenger, E. (1990). *Situated Learning: Legitimate Peripheral Participation*. Cambridge, UK: Cambridge University Press.

Leifer, R., & Mills, P. (1996). An Information Processing Approach for Deciding upon Control Strategies and Reducing Control Loss in Emerging Organizations. *Journal of Management, 22,* 113–137.

Leonard-Barton, D. (1992). Core Capabilities and Core Rigidities: A Paradox in Managing New Product Development. *Strategic Management Journal, 13,* 111–125.

Leonard-Barton, D. (1995). *Wellsprings of Knowledge: Building and Sustaining the Sources of Innovation*. Boston: Harvard Business School Press.

Levitt, B., & March, J. G. (1988). Organizational Learning. *Annual Review of Sociology, 14,* 319–338

Lipnack, J., & Stamps, J. (1997). *Virtual Teams People Working Across Boundaries with Technology*. New York: Wiley.

Malone, K., & Crowston, K.(1994). The interdisciplinary study of coordination. *ACM Computers Surveys, 26,* 87–119.

March, J. G. (1972). Model bias in social action. *Review of Educational Research, 42*(4), 413–429

March, J. G., & Simon, A. H. (1958). *Organizations*. New York: Wiley.

Martines, L., & Kambil, A. (1999). Looking Back and Thinking Ahead: Effects of Prior Success on Managers' Interpretations of New Information Technologies. *Academy of Management Journal, 42,* 652–661.

Marx, K. (1930). *Critique of Political Economy*. New York: Dutton.

Marx, K. (1967). *Writing of Young Marx on Philosophy and Society*. New York: Dutton.

Mata, F. J., Fuerst, W. L., & Barney, J. B. (1995). Information Technology and Sustained Competitive Advantage: A Resource-Based Analysis. *MIS Quarterly, 19*(4), 487–505.

Mintzberg, H. (1978). Patterns in Strategy Formation. *Management Science, 24,* 934–948.

Mintzberg, H., & Walters, J. (1985). Of Strategies Deliberate and Emergent. *Strategic Management Journal, 6,* 257–272.

Nakamura, K. (2005, September). Management innovation through IT (in Japanese). *Diamond Harvard Business Review*, pp. 51–52.

Nihon Keizai Shimbun. (1998, November). *Determination of the Japan Quality Award in 1998, 19,* 26.

Nikkei Information Strategy. (1998). *Case Studies of Video Conferences*, pp. 68, 69.

Nikkei Sangyo Shimbun. (1998, February). *Product Information Sending and Receiving System Configuration (IBIZA), 18,* 3

Nohria, N., & Ghoshal, S. (1997). *The Differentiated Network: Organizing Multinational Corporations for Value Creation.* San Francisco: Jossey-Bass.

Nolan, R. L. (1979). Managing the Crises in Data Processing. *Harvard Business Review, 57*(2), 115–126.

Nonaka, I., & Konno, N. (1998). The Concept of "Ba": Building a Foundation for Knowledge Creation. *California Management Review, 40,* 40–54.

Nonaka, I., & Takeuchi, H.(1995). *The Knowledge Creating Company.* Oxford, UK: Oxford University Press.

Ocker, R., Fjemestad, J., Hiltz, R., & Johnson, K. (1998). Effects of four models of group communication on the outcomes of software requirements determination. *Journal of Management Information Systems, 15*(1), 99–118.

Odenwald, S. (1996). *Global Solutions for Teams: Moving from Collision to collaboration.* Chicago: Irwin Professional Publishing.

O'Hara-Devereaux, M., & Johansen, R. (1994). *GlobalWork; Bridging Distance, Culture and Time.* San Francisco: Jossey-Bass.

Ohira, H. Kodama, M., & Yoshimoto, M. (2003). A World First Development of a Multipoint Videophone System over 3G–324M Protocol. *International Journal of Mobile Communications, 1*(3), 264–272.

Olikowski, W. J. (1992). The Duality of Technology: Rethinking the Concept of Technology in Organizations. *Organization Science, 3*(3), 398–427.

Olikowski, W. J. (1996). Improvising Organizational Transformation Over Time: A Situated Change Perspective. *Information Systems Research, 7*(2), 63–92.

Olikowski, W. J. (2000). Using Technology and Constituting Structures: A Practice Lens for Studying Technology in Organizations. *Organization Science, 11*(4), 404–428.

Olikowski, W. J., & Barley, S. R. (2001). Technology and Institutions: What Can Research on Information Technology and Research on Organizations Learn from Each Other?. *MIS Quarterly, 25,* 145–165.

Olikowski, W. J., & Robey, D. (1991). Information technology and the structuring of organizations, *Information Systems Research, 2*(2), 143–169.

Orr, J. (1996). *Talking about Machines: An Ethnography of a Modern Job.* Ithaca, NY: ILP Press.

Osterlof, M., & Frey, B. (2000). Motivation, Knowledge Transfer, and Organizational Forms. *Organization Science, 11*(3), 538–550.

Paul, J., Costley, L., Howell, P., & Dorfman, W. (2002). The Mutability of Charisma in Leadership Research. *Management Decision, 40*(1), 192–197.

Peltokorpi, V.M., Nonaka, I., & Kodama, M. (2007). NTT DoCoMo`s Launch of I-Mode in the Japanese Mobile Phone Market: A Knowledge Creation Perspective. *Journal of Management Studies, 44(1),* 50–72.

Peng, K., & Akutsu, S. (2001). A Mentality Theory of Knowledge Creation and Transfer: Why Some Smart People Resist New Ideas and Some Don't. In I. Nonaka & D. Teece (Eds.), *Managing Industrial Knowledge: Creation, Transfer and Utilization* (pp. 105–123). London: Sage.

Peng, K., & Nisbett, R. E. (1999). Culture Dialectics, and Reasoning about Contradiction. *American Psychologist, 54,* 741–754.

Polanyi, M. (1966). *The Tacit Dimension.* New York: Doubleday.

Politis, J. D. (2001). The Relationship of Various Leadership Styles to Knowledge Management. *Leadership and Organizational Development Journal, 22*(8), 354–364.

Popper, M., & Lipshitz, R. (2000). Installing Mechanisms and Instilling Values: The Role of Leaders in Organizational Learning. *The Learning Organization, 7*(3), 135–145.

Porter, M. (1985). *Competitive Advantage.* New York: Free Press.

Porter, M. E. (1991). Toward a Dynamic Theory of Strategy. *Strategic Management Journal, 12,* 95–117.

Porter, M. E. (2001, March). Strategy and the Internet. *Harvard Business Review,* pp. 63–78.

Powell, T. C., & Dent-Micallef, A. (1997). Information technology as competitive advantage: the role of human, business, and technology resources. *Strategic Management Journal, 18*(5), 375–405.

Powell, W., Koput, K., & Smith-Doerr, L. (1996). Inter-Organizational Collaboration and the Locus of Innovation: Networks of Learning in Biotechnology. *Administrative Science Quarterly, 41,* 116–146.

Prahalad, C. K., & Hammel, G. (1990, May-June). The Core Competence of the Corporation. *Harvard Business Review,* pp. 79–91.

Ring, S., & Van de Ven, A. (1994). Developmental Process of Cooperative Interorganizational Relationships. *Academy of Management Review, 23*(3), 393–404.

Robey D., Boudreau M.-C., & Rose G. M. (2000). Information Technology and Organizational Learning: A Review and Assessment of Research. *Accounting, Management and Information Technologies, 10*(2), 125–155.

Robey, D., & Sahay, S. (1996). Transforming Work Through Information Technology: A Comparative Case Study of Geographic Information Systems in County Government. *Information Systems Research, 7,* 93–110.

Rosenberg, N. (1982). *Inside the Black Box: Technology and Economics.* Cambridge, UK: Cambridge University Press.

Ross, J. W., Beath, C. M., & Goodfue, D. L. (1996). Building Long-Term Competitiveness through IT Assets. *Sloan Management Review, 38*(1), 31–45.

Sadler, P. (2001). Leadership and Organizational Learning. In M. Dierkes, et al. (Eds.), *Handbook of Organizational Learning and Knowledge* (pp. 415–442). Oxford, UK: Oxford University Press.

Sakai, H., & Amasaka, K. (2005). V-MICS, Advanced TPS for Strategic Production Administration: Innovative Maintenance Combining DB And CG. *Journal of Advanced Manufacturing Systems, 4*(1). 5–20.

Schank, R., & Abelson, C. R. (1977). Scripts, Plans, and Goods. Hillsdale, NJ: Erlbaum.

Schon, A. (1987). *Educating the Reflective Practitioner.* San Francisco: Jossey-Bass.

Seifter, H., & Economy, P. (2001). *Leadership Ensemble: Lessons in Collaborative Management from the World's Only Conductorless Orchestra.* New York: Times Books.

Seo, M., & Douglas Creed, W. (2002).Institutional Contradictions, Praxis, and Institutional Change: A Dialectical Perspective. *Academy of Management Review, 27*(2), 222–247.

Shannon, C., & Weaver, W. (1949). *The Mathematical Theory of Communications.* Urbana: University of Illinois Press.

Shapiro, C., & Varian, H. R. (1998). *Information Rules*. Boston: Harvard Business School Press.

Shiga, T. (2006, January). Negotiation through videoconferencing. *Nihon keizai Shinbun, 16*, 4.

Shone, D. A. (1983). *The Reflective Practitioner: How Professionals Think in Action*. New York: Basic Books.

Skyrrme, D. J. (2001). *Capitalizing on Knowledge: From E-business to K-business*. London: Butterworth-Heinemann.

Spears, L. (1995). *Reflections on Leadership*. New York: Wiley.

Spender, J. C. (1990). *Industry Recipes: An Enquiry into the Nature and Sources of Managerial Judgement*.Oxford, UK: Blackwell.

Spender, J. C. (1992). Knowledge Management: Putting your Technology Strategy on Track. In T. M. Khalil & B. A. Bayraktar (Eds.), *Management of Technology* (Vol. 3, pp. 404–413). Norcross, GA: Industrial Engineering and Management Press.

Stalk, G., Evans, P., & Schulman, L. E. (1992, March-April). Competing on Capabilities: The New Rules of Corporate Strategy. *Harvard Business Review*, pp. 57–69.

Star, S. L. (1989). The Structure of Ill-Structured Solutions: Boundary Objects and Heterogeneous Distributed Problem Solving. In M. Huhns & I. L. Gasser (Eds.), *Readings in Distributed Artificial Intelligence*. Menlo Park, CA: Morgan Kaufman.

Suchman, L. (1987). *Plans and Situational Actions*. Cambridge, UK: Cambridge University Press.

Szulanski, G. (2000). The process of knowledge transfer: A diachronic analysis of stickiness. *Organization Behavior Human Decision Processes, 82*, 9–27.

Teece, D.J., Pisano, G., & Shuen, A. (1997). Dynamic Capabilities and Strategic Management. *Strategic Management Journal, 18*(3), 509–533.

Tichy, N. M., & Devanna, M. (1986). *The Transformational Leader*. Chichester, UK: Wiley.

Tippens, M. J., & Sohi, R. S. (2003). IT Competency and Firm Performance: Is Organizational Learning a Missing Link?.*Strategic Management Journal, 24*(8), 745–761.

Toffler, A. (1990). *Powershift: Knowledge, Wealth and Violence at the Edge of the 21st Century*. New York: Bantam Books.

Tokachi Mainichi News. (1999a). *Animal Medical Information and Science Development Research Institute: 25 points nation-wide connected to video phone net conference*. February 1, 1999.

Tokachi Mainichi News. (1999b). *Bidirectional Data Networks in Veterinary Medicine: Construction of Remote Medical Treatment System*. January 1, 1999.

Tokachi Mainichi News. (1999c). *Gigabit Network Practical Research: Project Adopted from Obihiro* December 4, 1999.

Tsoukas, H. (1996). The firm as a distributed knowledge system: A constructionist approach. *Strategic Management Journal, 17*(1), 11–25.

Van Maanen, J., & Yates, J. (2001). *Information Technology and Organizational Transformation: History, Rhetoric and Preface*. London: Sage.

Van de Ven, A. H., & Poole, M. S. (1995). Explaining Development and Change in Organizations. *Academy of Management Review, 20*(5), 510–540.

Venkatraman, N. (1991). IT induced Business Reconfiguration. In M. S. Morton (Ed.), *The Corporation of the 1990s: Information Technology and Organizational Transformation*. Oxford, UK: Oxford University Press.

Venkatraman, N. (1994). IT-Enabled Business Transformation: From Automation to Business Scope Redefinition. *Sloan Management Review, 34*(2), 73–87.

Venkatraman, N. (1997). Beyond Outsorcing: Managing IT Resources As a Value Center. *Sloan Management Review, 38*(3), 51–64.

Walker, D. M., Walker, T., & Schmitz, R.(2002). *Doing Business Internationally: The Guide to Cross-Cultural Success*. New York: McGraw-Hill.

Walsh, J., & Urgson, G. (1991). Organizational memory. *Academy of Management Review, 16*, 57–91.

Walsham, G. (1997). Actor-network theory and IS research:current status and future prospects. In *Proceedings of the IFIP TC8 WG 8.2 international conference on Information systems and qualitative research* (pp. 466–480). New York: Chapman & Hall.

Watkins, K., & Marsick, V. (1993) *Sculpting the Learning Organization*. San Francisco: Jossey-Bass.

Weick, K. E. (1979). *The Social Psychology of Organizing* (2nd ed.). Reading, MA: Addison-Wesley.

Weick, K E. (1995). *Sensemaking in Organizations*. London: Sage.

Weick, K. E., & Browning, L. (1986). Argument and Narration in Organizational Communication. *Journal of Management, 12*(2), 243–259.

Weill, P., & Broadbent, M. (1998). *Leveraging the New Infrastructure: How Market Leaders Capitalize on Information Technology*. Boston: Harvard Business School Press.

Weill, P., & Ross, J. W. (2004). *It Governance: How Top Performers Manage It Decision Rights for Superior Results*. Boston: Harvard Business School Press.

Wenger, E. C. (1998). *Community of Practice: Learning, Meaning and Identity*. Cambridge, UK: Cambridge University Press.

Wenger, E. C. (2000). Communities of Practice: The Organizational Frontier. *Harvard Business Review, 78*(1), 139–145.

Wertsch, J. V. (1991). *Voice of the Mind: A Sociocultural Approach to Mediated Action*. Cambridge, MA: Harvard University Press.

Wertsch, J. V. (2000). Intersubjectivity and Alterity in Human Communication. In N. Budwig, I. C. Uzgiris, & J. V. Wertsch (Eds.), *Communication: An Arena of Development* (Vol. 19). Stamford, CT: Ablex Publishing.

Westley, F., & Vredenburg, H. (1991). Strategic Bridging: The Collaboration between Environmentalists and Business in the Marketing of Green Products. *Journal of Applied Behavioral Science, 27*(2), 65–90.

Wisemann, C. (1988). *Strategic Information Systems*. New York: Richard D. Irwin.

Womack, J., & Jones, D. (1996). *Lean Thinking: Banish Waste and Create Wealth in Your Corporation*. New York: Simon & Schuster.

Wood, D. J., & Gray, B. (1991). Toward a Comprehensive Theory of Collaboration. *Journal of Applied Behavioral Sciences, 27*(2), 139–163.

Zigurs, I. (2003). Leadership in Virtual Teams: Oxymoron or Opportunity?. *Organizational Dynamics, 31*(4), 339–351.